D1240109

Beginning SQL Server Reporting Services

Kathi Kellenberger

Apress®

Beginning SQL Server Reporting Services

Kathi Kellenberger
Edwardsville, Illinois
USA

ISBN-13 (pbk): 978-1-4842-1989-8   ISBN-13 (electronic): 978-1-4842-1990-4
DOI 10.1007/978-1-4842-1990-4

Library of Congress Control Number: 2016951945

Copyright © 2016 by Kathi Kellenberger

This work is subject to copyright. All rights are reserved by the Publisher, whether the whole or part of the material is concerned, specifically the rights of translation, reprinting, reuse of illustrations, recitation, broadcasting, reproduction on microfilms or in any other physical way, and transmission or information storage and retrieval, electronic adaptation, computer software, or by similar or dissimilar methodology now known or hereafter developed.

Trademarked names, logos, and images may appear in this book. Rather than use a trademark symbol with every occurrence of a trademarked name, logo, or image we use the names, logos, and images only in an editorial fashion and to the benefit of the trademark owner, with no intention of infringement of the trademark.

The use in this publication of trade names, trademarks, service marks, and similar terms, even if they are not identified as such, is not to be taken as an expression of opinion as to whether or not they are subject to proprietary rights.

While the advice and information in this book are believed to be true and accurate at the date of publication, neither the authors nor the editors nor the publisher can accept any legal responsibility for any errors or omissions that may be made. The publisher makes no warranty, express or implied, with respect to the material contained herein.

Managing Director: Welmoed Spahr
Lead Editor: Jonathan Gennick
Technical Reviewer: Rodney Landrum
Editorial Board: Steve Anglin, Pramila Balan, Laura Berendson, Aaron Black, Louise Corrigan, Jonathan Gennick, Robert Hutchinson, Celestin Suresh John, Nikhil Karkal, James Markham, Susan McDermott, Matthew Moodie, Natalie Pao, Gwenan Spearing
Coordinating Editor: Jill Balzano
Copy Editor: Lori Jacobs
Compositor: SPi Global
Indexer: SPi Global
Artist: SPi Global
Cover Image: Designed by Freepik.com

Distributed to the book trade worldwide by Springer Science+Business Media New York, 233 Spring Street, 6th Floor, New York, NY 10013. Phone 1-800-SPRINGER, fax (201) 348-4505, e-mail orders-ny@springer-sbm.com, or visit www.springer.com. Apress Media, LLC is a California LLC and the sole member (owner) is Springer Science + Business Media Finance Inc (SSBM Finance Inc). SSBM Finance Inc is a Delaware corporation.

For information on translations, please e-mail rights@apress.com, or visit www.apress.com.

Apress and friends of ED books may be purchased in bulk for academic, corporate, or promotional use. eBook versions and licenses are also available for most titles. For more information, reference our Special Bulk Sales–eBook Licensing web page at www.apress.com/bulk-sales.

Any source code or other supplementary material referenced by the author in this text is available to readers at www.apress.com. For detailed information about how to locate your book's source code, go to www.apress.com/source-code/.

Printed on acid-free paper

For Nate. I love your handsome little face!

Contents at a Glance

Contents

About the Author

Kathi Kellenberger known to the SQL Server community as Aunt Kathi, is an independent SQL Server consultant associated with Linchpin People and a Data Platform MVP. She loves writing about SQL Server and has contributed to more than a dozen books as an author, coauthor, or technical reviewer. Kathi enjoys spending free time with family and friends, especially her five grandchildren. When she is not working or involved in a game of Hide 'n Seek with the kids, you may find her at the local Karaoke bar. Kathi's blog can be found at `www.auntkathisql.com`.

About the Technical Reviewer

Rodney Landrum went to school to be a poet and a writer. And then he graduated, so that dream was crushed. He followed another path, which was to become a professional in the fun-filled world of information technology. He has worked as a systems engineer, UNIX and network administrator, data analyst, client services director, and finally database administrator. The old hankering to put words on paper, while paper still existed, got the best of him, and in 2000, he began writing technical articles—some creative and humorous, some quite the opposite. In 2009 he wrote *The SQL Server Tacklebox* (Simple Talk Publishing, 2009), a title his editor disdained but a book closest to the true creative potential he sought; he wanted to do a full book without a single screenshot. He promises his next book will be fiction or a collection of poetry, but that has yet to transpire.

Acknowledgments

Once again, I have placed my words inside a book. For me, writing really is a labor of love. The best thing in the world is to be approached at a conference by someone who has read one of my books and learned a new skill because of it. This book is for all of you out there who have thanked me for writing, letting me know that I have made a difference in your career if not your life.

Of course, I must thank my husband and family for putting up with me saying so often, "I can't—I need to work on my book." I hope I have managed to be there for you when you really needed me.

Thanks to Rodney, Jonathan, and Jill for helping me to get the book to completion in great shape. Thanks to Microsoft for giving SQL Server Reporting Services the love it deserves in the 2016 release.

And, thank you to everyone who reads this book. I hope you enjoy working with SSRS as much as I have over the years.

PART I

Getting Started

CHAPTER 1

Getting Started

At my first job as a database administrator, I was asked to look at a problem with some reports. The reports were created in MS Access and linked to a SQL Server database. Each manager had his or her own version of the reports and, even though the reports had started out the same at one time, they had been modified by the individual managers over the years. The managers were complaining that the numbers were not consistent, and could I fix the problem?

I worked to correct the discrepancies as best I could, but the individual copies of reports still existed. Shortly after this, I attended the 2003 PASS Summit and saw the announcement about SQL Server Reporting Services (SSRS). In 2004, Microsoft released SSRS as an add-in for SQL Server 2000. I didn't wait for the release. I knew that SSRS was going to solve my MS Access report problem, and I installed SSRS as soon as it was available.

The advantage that SSRS brought, compared to the MS Access reports, was the centralized web site, Report Manager, where the reports were published. Instead of each manager having his or her own copy of reports, the managers would run the reports from a central location eliminating the discrepancies.

SQL Server Reporting Services is one of the core components of the Microsoft Business Intelligence stack. SSRS is a feature-rich reporting tool that now includes mobile reports as well as a modern on-premises web portal.

SSRS has a number of interactive features, visual elements such as charts and maps, security, and more. Reports can contain data displayed in tabular format or visually. You can also create attractive and informative dashboards.

To run reports, end users browse to the web portal and click the report name. In the background, SSRS requests the data from the source databases and builds the report. The report is then delivered to the end user. Figure 1-1 shows how this works.

Electronic supplementary material The online version of this chapter (doi:10.1007/978-1-4842-1990-4_1) contains supplementary material, which is available to authorized users.

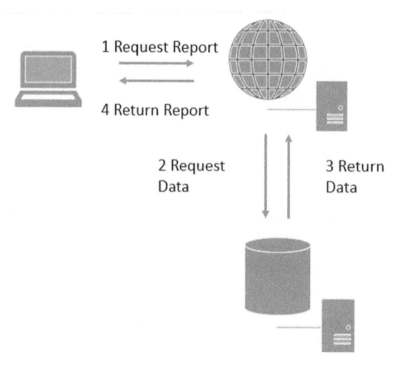

Figure 1-1. *Reporting steps*

Understanding SSRS Architecture

An SSRS implementation consists of multiple components that can be configured in many different ways. At a minimum, everything can go on one computer, even on a laptop. This configuration is probably useful only for development, and it is what I recommend for following along with the examples in this book. The configuration consists of a SQL Server instance that includes Reporting Services as well as the source databases and SQL Server Data Tools (SSDT) running in Visual Studio.

▓ **Note**   SSRS can also be installed in SharePoint integrated mode. The way you develop reports is identical to the default which is called native mode. This book will focus on native mode, but it does have a section in Chapter 8 on deploying reports to SharePoint.

Typically, in a production environment, a server is dedicated to running SSRS, and the source data is found on other servers throughout the network. Report developers will use SSDT on their local computers to develop the reports and then publish the reports to the production server or possibly to a server where the reports can be tested before going live.

Before learning how to get everything set up on your computer, you will learn more about the components of SSRS. First there must be a SQL Server instance in place to host the SSRS databases. The instance is often installed on the server where the SSRS service is installed, but it can be a different server.

There are two databases that will be created when you install or initially configure SSRS: ReportServer and ReportServerTempDB. ReportServer is used to store report definitions, security, history, and everything else that is needed for the published reports. You can probably tell by the name ReportServerTempDB that this database is used as a temporary workspace.

When you install SSRS, it creates a web service that responds to report requests. In native mode, it provides a web portal where users can browse for and run reports. In previous versions of SSRS this was called Report Manager, but starting with 2016 this interface has been completely redesigned. It is now just called the web portal and resembles Figure 1-2.

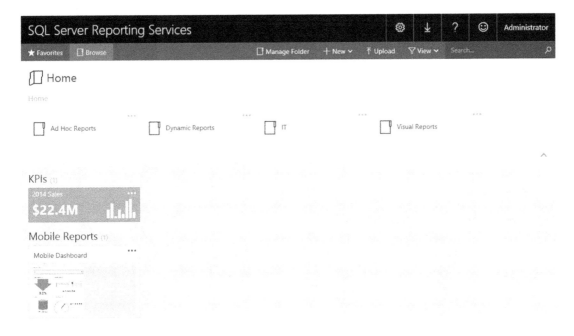

Figure 1-2. *The web portal*

The source of data can be from just about anywhere. This book will show examples from SQL Server databases, but you could report against Oracle, Analysis Services cubes, XML documents, SharePoint lists, cloud databases, and more.

Installing SQL Server with SSRS

You can follow along with many of the examples in the book by installing the developer tools without installing SSRS. You could also work with an SSRS instance that is already in your company's network. I do recommend that, if at all possible, you install SSRS on your development computer. That will allow you to learn how to do some administrative tasks as well as develop the reports.

SQL Server is available in several editions. Each edition has a specific set of features and price. For development and learning, you can download the free Developer Edition. Just search the web for "SQL Server Developer Edition download" to find the file. There is also an Express Edition that is free, but the features are very limited.

■ **Note** At the time of this writing, the media is an iso file. My Windows 10 laptop can easily handle iso files, but your operating system may not. You can search for a utility to mount or extract iso files if needed.

From the media, you should see a setup icon shown in Figure 1-3.

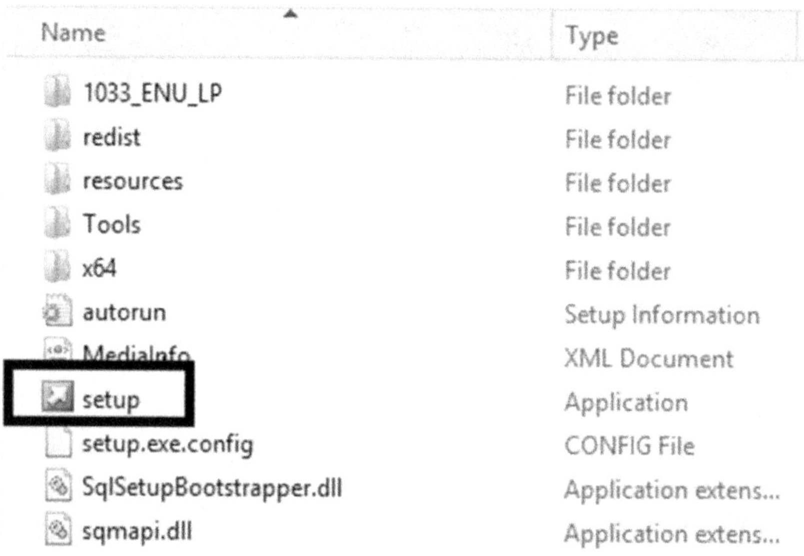

Name	Type
1033_ENU_LP	File folder
redist	File folder
resources	File folder
Tools	File folder
x64	File folder
autorun	Setup Information
MediaInfo	XML Document
setup	Application
setup.exe.config	CONFIG File
SqlSetupBootstrapper.dll	Application extens...
sqmapi.dll	Application extens...

Figure 1-3. *The setup icon*

Follow these instructions to install a SQL Server instance with SSRS:

1. Double-click setup to launch the SQL Server Installation Center.

2. Click Installation on the left.

3. Click New SQL Server stand-alone installation or add features to an existing installation at the top as shown in Figure 1-4.

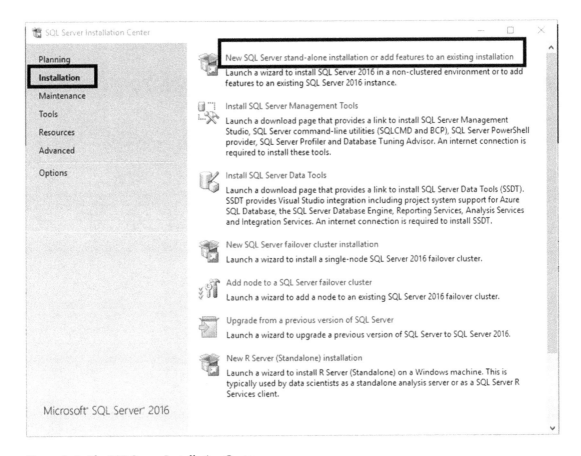

Figure 1-4. *The SQL Server Installation Center*

4. An installation wizard will launch. On the initial information pages, click Next.

5. On the License Terms page, click I accept the License Terms and click Next.

6. Click Next on the Microsoft Update page.

7. After checking for updates, click Next on the Product Updates.

8. On the Install Rules page, click Next once it is done. If there are any Failed statuses, you will need to click the message to find out what is wrong and correct it.

9. On the Feature Selection page, select Database Engine Services and Reporting Services – Native as shown in Figure 1-5.

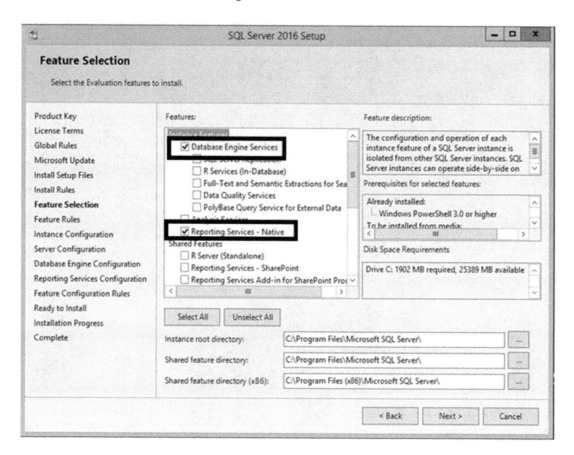

Figure 1-5. *The feature selection*

10. On the Instance Configuration page, you must decide whether to install a default instance with Instance ID MSSQLSERVER or a named instance. Each instance of SQL Server on a computer must be unique. If there are existing instances of SQL Server installed, you will see them listed. If no other default instance is installed, select Default Instance and click Next. Otherwise, select Named Instance and type in a name before clicking Next. Figure 1-6 shows this page.

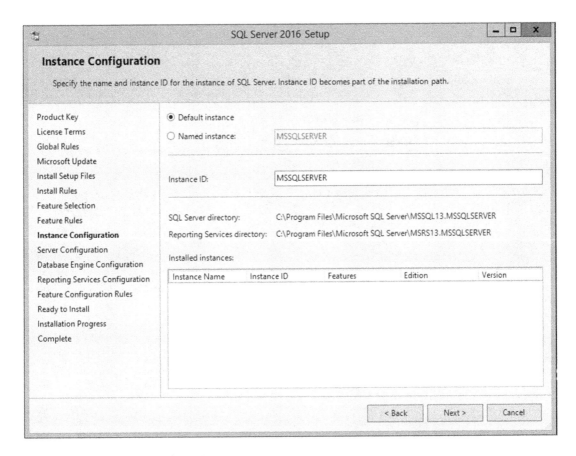

Figure 1-6. *The Instance Configuration page*

11. On the Server Configuration page, accept the defaults and click Next.

12. On the Database Engine Configuration page, click Add Current User. This will make your account an administrator in SQL Server. Click Next.

13. On the Reporting Services Configuration page, make sure that you choose Install and configure as shown in in Figure 1-7 and click Next.

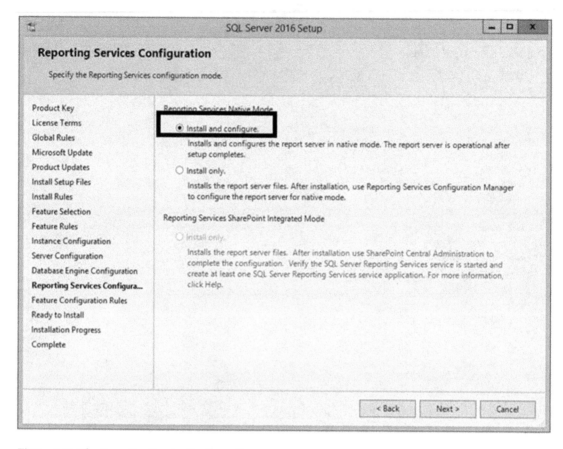

Figure 1-7. *The Reporting Service Configuration page*

14. On the Ready to Install page, click Install.

15. Restart the computer if requested to do so at the end of the installation.

It may take several minutes to install the SQL Server instance and SSRS. There are dozens of things that could prevent a successful installation, and it would be impossible for me to help you troubleshoot via a book. My advice is to navigate to C:\Program Files\Microsoft SQL Server\130\Setup Bootstrap\Log. There will be log files with the messages generated during the installation. You can search the Internet using any error messages that you find for help and advice if the installation fails. That said, you may need to be connected to the Internet during the installation, and you may need to run the setup as an administrator for a successful install.

Previous versions of SQL Server allowed you to install SQL Server Management Studio (SSMS) with your SQL Server instance install. Starting with SQL Server 2016, Microsoft plans to release frequent updates to this tool, and make it available only by downloading. To find the link, relaunch the SQL Server Installation Center if you have closed it. On the Installation page, click Install SQL Server Management Tools. Follow the instructions found on the download page.

Installing SQL Server Data Tools

The primary development tool for SSRS is SSDT, mentioned earlier, and it runs inside Visual Studio. Microsoft has changed the name and the source of the development tool over several versions of SQL Server. At one time, you could install Business Intelligence Development Studio, also known as BIDS, directly from the SQL Server installation media. At one point, Microsoft changed the name to SQL Server Data Tools – BI, and it was a separate download. To make things confusing, there was also another product called SSDT used for database projects, not BI projects like reports. Luckily, in 2016, Microsoft has combined the two products into one SSDT download.

You can find the link to download and install SSDT on the Installation page of the SQL Server Installation Server as shown in Figure 1-8.

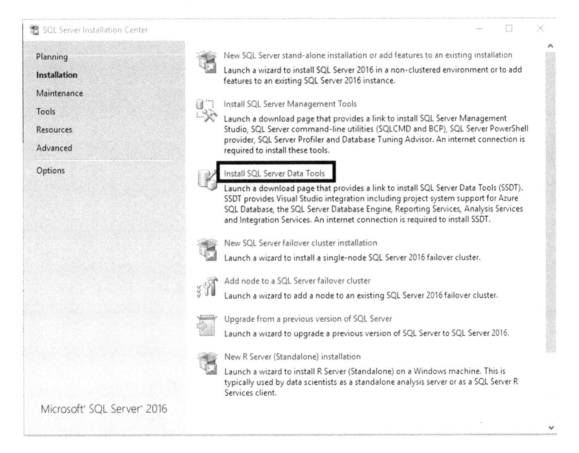

Figure 1-8. *The link for SQL Server Data Tools*

At the time of this writing, you can download the SSDTSetup.exe file and install from that, or you can scroll down the page to download an iso file. If you download the iso file, then run SSDTSetup.exe from the media to get the install started. Follow these steps to install SSDT:

1. Running SSDTSetup.exe starts the wizard. On the first page, make sure that SQL Server Reporting Services is checked as shown in Figure 1-9. You can leave the others checked as well.

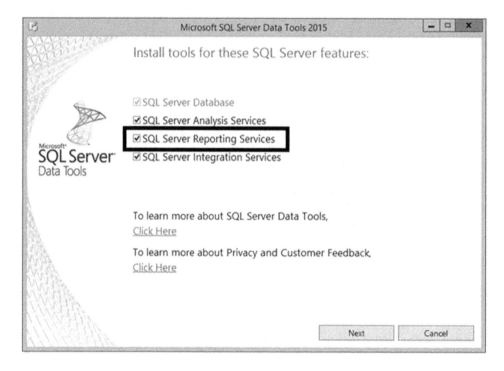

Figure 1-9. *SQL Server Reporting Services is checked*

2. Click Next.

3. On the License Terms page, check I agree to the license terms and conditions.

4. Click Install

Configuring SSRS

If you followed the installation instructions exactly in the section "Installing SQL Server with SSRS", SSRS should be configured. If, instead, you added SSRS to an existing SQL Server instance or selected Install only on the Reporting Services Configuration page, you will need to configure it now. To configure SSRS, follow these steps:

1. Launch Reporting Services Configuration Manager.

2. When asked to connect to your SSRS instance, select the server and instance name if required and click Connect as shown in Figure 1-10.

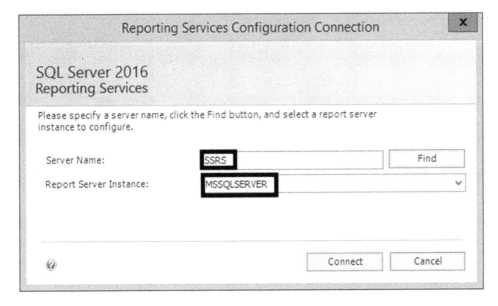

Figure 1-10. *Connect to the SSRS instance*

3. Select the Database page and click Change Database as shown in Figure 1-11.

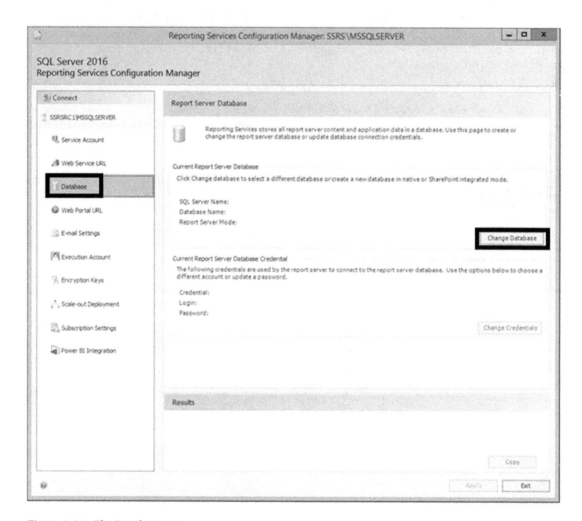

Figure 1-11. *The Database page*

4. This opens the Report Server Database Configuration Wizard. Select Create a new report server database as shown in Figure 1-12. Click Next.

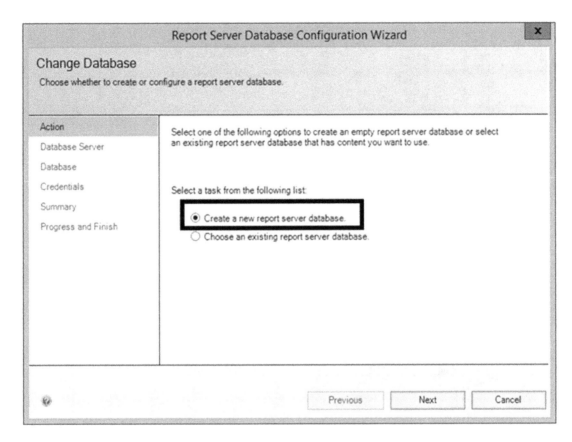

Figure 1-12. *Create a new report server database*

5. On the Database Server page, make sure that your server name is filled in. If you have a named instance, be sure to include the instance name. Figure 1-13 shows this page.

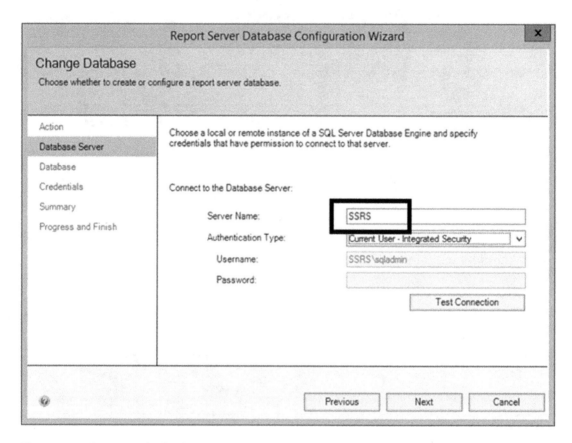

Figure 1-13. *Connect to the database server*

6. Click Next to move to the Database page shown in Figure 1-14. Accept the defaults on this page. If your installation is a named instance, the instance name will be part of the database name.

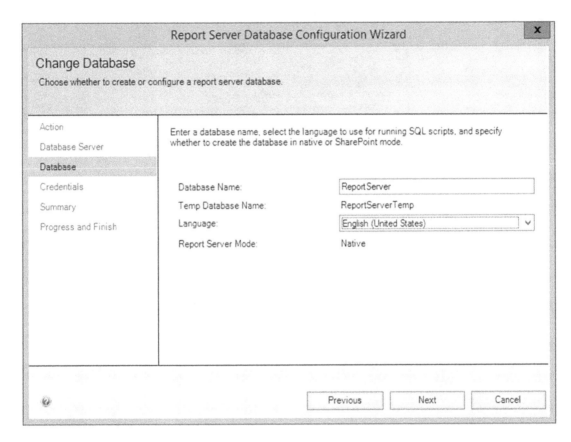

Figure 1-14. *The Database Name*

7. Click Next to move to the Credentials page. You can change the SSRS service account on this page. Accept the defaults and click Next.

8. Click through the remaining pages in the wizard to create the SSRS databases.

9. Click Finish once the process is complete.

10. To create the Web Service URL, select the Web Service URL page as shown in Figure 1-15.

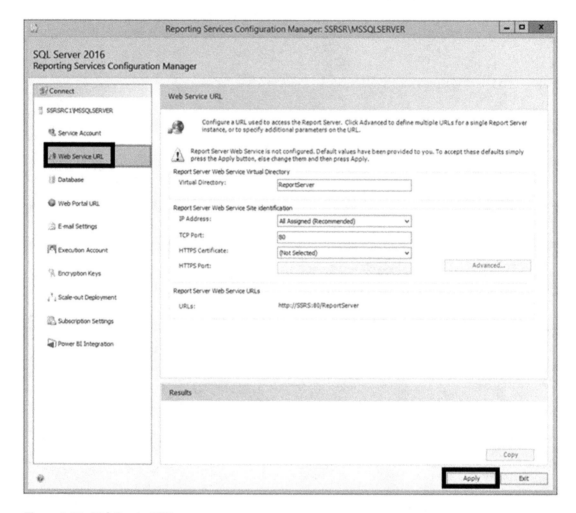

Figure 1-15. *Web Service URL page*

11. For your own SSRS installation, just accept the defaults and click Apply. This sets up the web service.

12. When the task has completed, select the Web Portal URL page. Once again, you can accept the defaults and click Apply. This will create the web portal.

13. When the web portal creation is done, click the Encryption Keys page as shown in Figure 1-16. Click Backup to save the encryption keys.

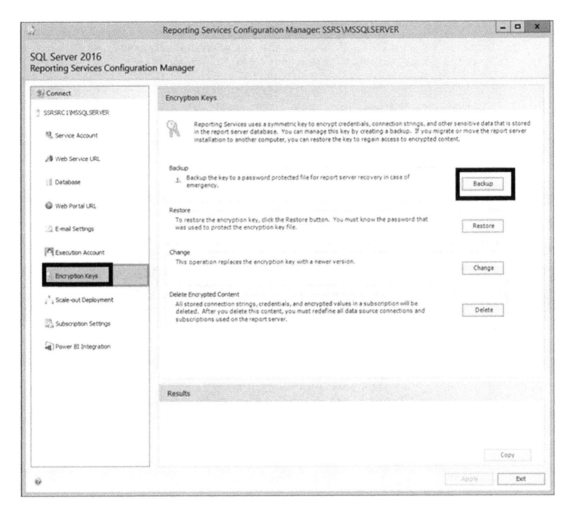

Figure 1-16. *Back up the encryption keys*

14. Supply a location and password that you will not forget. This step is especially important in a production environment. The encryption key is required for restoring or moving the database.

15. Click Exit to close the SSRS Configuration Manager.

SSRS should now be configured. In Chapter 8, you will learn how to publish your reports. At that time, you will return to this tool to determine the web service URL and web portal URL.

Configuring Local SSRS Settings

There is one very frustrating problem that you will encounter if you install the SSRS instance locally related to security. In order to launch the web portal or publish reports, you will need to run the web browser and SSDT as an administrator. This feature helps prevent applications from making changes to the operating system without your knowledge and permission.

To get around this issue, follow these steps:

1. Determine the web portal URL by launching Reporting Service Configuration Manager. Click the Web Portal URL page and note the link. Do not click it.

2. Launch your web browser using the Run as an Administrator option.

3. Navigate to the URL determined in step 1.

4. Open the security settings of the web browser and add the current site to the Trusted Sites.

5. Click OK and close the browser.

6. Launch the browser again using the Run as an Administrator option.

7. Navigate to the web portal URL once again.

8. Click Manage Folder as shown in Figure 1-17.

Figure 1-17. *The Manage Folder link*

9. Click Add group or user.

10. Type in your computer or domain plus the account as the Group or user.

11. Select Content Manager as the role. The dialog will look similar to that in Figure 1-18.

Use this page to define role-based security for **Home**.

Group or user: KathisPC\Kathi

Select one or more roles to assign to the group or user.

	Role	Description
☐	Browser	May view folders, reports and subscribe to reports.
☑	Content Manager	May manage content in the Report Server. This includes folders, reports and resources.
☐	My Reports	May publish reports and linked reports; manage folders, reports and resources in a users My Reports folder.
☐	Publisher	May publish reports and linked reports to the Report Server.
☐	Report Builder	May view report definitions.

| OK | Cancel |

Figure 1-18. *The security for Home*

12. Click OK.

13. Click the gear icon found at the top right of the page and select Site Settings as shown in Figure 1-19.

Figure 1-19. *The Site Settings link*

14. Select the Security page.

15. Click Add group or user.

16. Enter your account name and click System Administrator. The dialog will look like that in Figure 1-20.

Use this page to assign a user or group to a system role. You can also use this page to create or modify a system role definition.

Group or user: KathisPC\Kathi

Select one or more roles to assign to the group or user.

	Role	Description
☑	System Administrator	View and modify system role assignments, system role definitions, system properties, and shared schedules.
☐	System User	View system properties, shared schedules, and allow use of Report Builder or other clients that execute report definitions.

[OK]   [Cancel]

Figure 1-20. *The site settings*

17. Click OK.

You will learn more about these security settings in Chapter 9. If you have followed these instructions but still encounter security errors when launching the web portal or publishing reports, see the article found at the following site for more information: https://msdn.microsoft.com/en-us/library/bb630430.aspx.

Determining the SQL Server Name

To follow the examples in this book, you will need to connect to your SQL Server instance when creating data sources. A SQL Server name has two parts: a computer name and an instance name. For example, a SQL Server located on a server named MyServer and an instance name of Inst1 can be reached with MyServer\Inst1. Often, SQL Server will be installed as the default instance. In that case, you do not need to supply the instance name which is actually MSSQLSERVER; you can supply just the computer name.

In the section "Installing SQL Server with SSRS," you were instructed to install SQL Server as the default instance. If you supplied an instance name instead, you will need that name to connect to the database. To find out the instance name, you will need to launch SQL Server Configuration Manager.

■ **Note**   If you are using a network SQL Server instead of a locally installed instance, ask the person who is responsible for that server for the correct computer and instance names.

Once Configuration Manager is running, click SQL Server Services. If you see MSSQLSERVER, you have a default instance. If you see anything else, that is your instance name. Figure 1-21 shows both a default instance and a named instance called SSRS.

Figure 1-21.   *The SQL Server Configuration Manager*

The SQL Server Configuration Manager utility has many other uses that are beyond the scope of this book.

Restoring the AdventureWorks Database

There are quite a few pieces to get into place in order to follow the examples in this book. The last is the AdventureWorks database which is often used as a sample database for SQL Server. Because the download locations change from time to time, browse to www.codeplex.com and then search for Microsoft SQL Server Product Samples: Databases. On that page, you may see links for several different versions. Make sure you find the download for AdventureWorks2016.bak.

■ **Note**   At the release of SQL Server 2016, the CodePlex page had an AdventureWorks2016CTP3.bak file available but not an AdventureWorks2016.bak file. The CTP3 designation refers to Community Technology Preview 3, a Beta version of SQL Server. If the release version is not available, the CTP3 version will work, but you will need to change the database name during the restore process.

Follow these steps to restore the database:

1. Download the AdventureWorks2016.bak or AdventureWorks2016CTP3.bak file.

2. The downloaded file most likely ended up in your Download folder. In order to restore the file, you will need to move it. If C:\temp doesn't exist, create it. Move the file to the new folder.

3. Launch SQL Server Management Studio.

4. Use the server name you determined in the previous section when prompted to connect. If you installed SQL Server locally, you can use localhost, (local), or a period as shown in Figure 1-22.

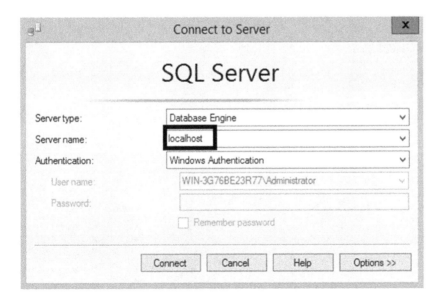

Figure 1-22. *Connect to Server dialog*

5. Click the Connect button.

6. In the Object Explorer, right-click on Databases and select Restore Database as shown in Figure 1-23.

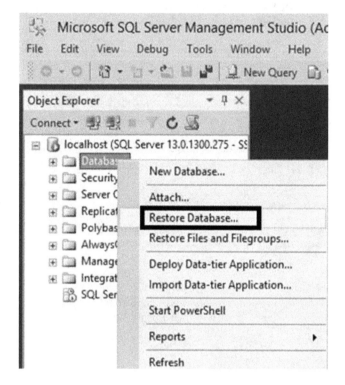

Figure 1-23. *The Restore Database selection*

7. On the Restore Databases dialog, select Device.

8. Click the ellipsis as shown in Figure 1-24.

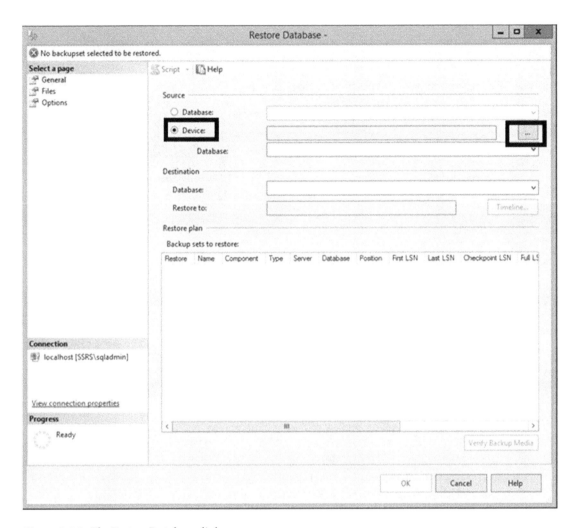

Figure 1-24. *The Restore Database dialog*

9. On the Select backup devices dialog, click Add and then navigate to the file as shown in Figure 1-25.

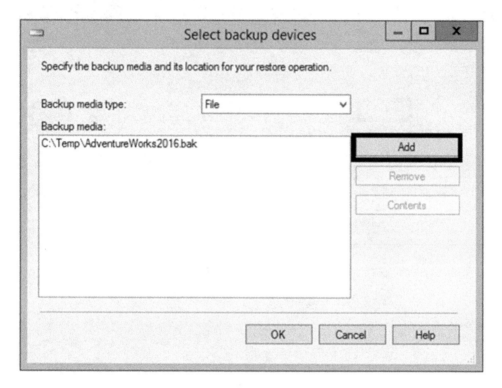

Figure 1-25. *The Select backup devices dialog*

10. Click OK to accept the file.

11. If the CTP3 file is the only one available for download, change the Database property from AdventureWorks2016CTP3 to AdventureWorks2016.

12. Click OK to start the restore.

13. Once the restore is complete click OK to dismiss the restore utility.

14. Expand and refresh the Databases folder. You should be able to see the AdventureWorks2016 database in place as shown in Figure 1-26.

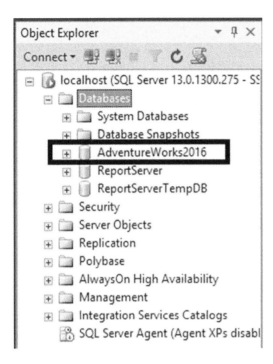

Figure 1-26. *The new database*

15. Exit SSMS.

In this case, the source of data and the SSRS databases are hosted on the same SQL Server instance. In most production environments, they would be hosted on separate servers.

Taking a Tour of SSDT

You will be spending quite a bit of time using SSDT throughout many of the chapters of this book. Take some time now to become familiar with it by following these steps:

1. Launch Visual Studio 2015.

2. Since this is the first time that it is launched, you must configure a couple of settings. Select Business Intelligence Settings.

3. Choose a theme that you prefer as shown in Figure 1-27.

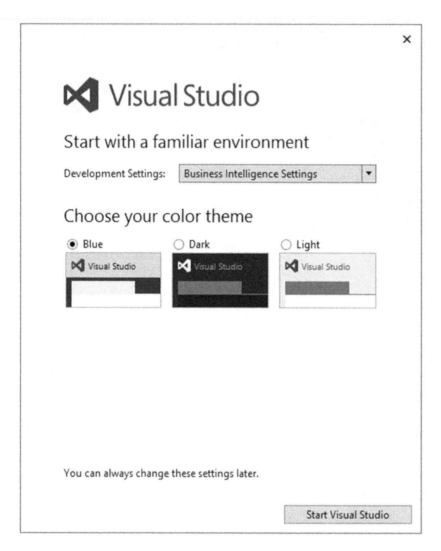

Figure 1-27. *Setting up the environment*

4. Click Start Visual Studio

5. Once Visual Studio is running select File ➤ New ➤ Project as shown in Figure 1-28.

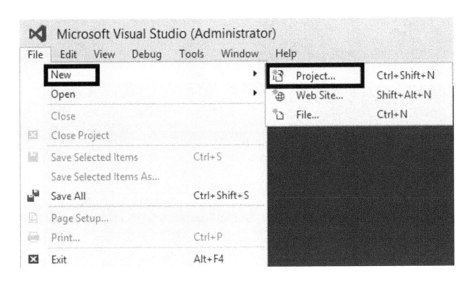

Figure 1-28. *Create a new project*

6. On the New Project dialog, select Report Server Project as shown in Figure 1-29.

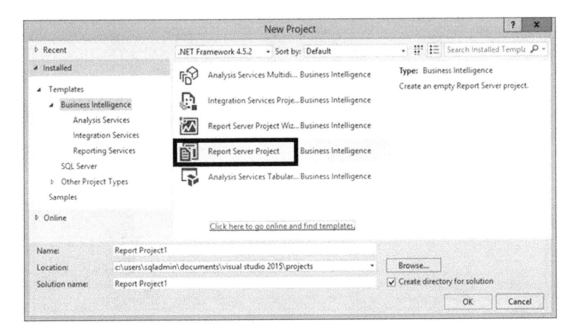

Figure 1-29. *Select Report Server Project*

7. Click OK to create the project.

There are several windows that you will use while developing reports. You'll learn all about them starting in Chapter 2. For now, if you are not familiar with Visual Studio, take some time to learn how the windows work. Each window can be repositioned, auto-hidden, or closed. Figure 1-30 shows the icons found at the top of each window.

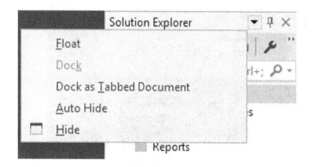

Figure 1-30. The window icons

The Auto Hide feature that you can also enable by clicking the pin icon hides the window without actually closing it. You can see the title on the edge of the program, and, by clicking the title, the window opens just when you need it. That's a nice feature to give you more room to work.

By clicking and dragging the window titles, you can move windows around. To see where they will end up, pay attention to the markers as you mouse over them and the areas that are highlighted as shown in Figure 1-31. The window will end up in the highlighted area when you drop the window.

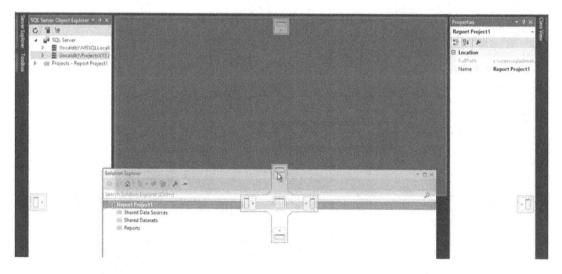

Figure 1-31. Moving the windows

If you close a window, you can always get it back by looking in the View menu. If you decide to go back to the default configuration, click Window ➤ Reset Window Layout.

When you close out of Visual Studio, the next time you open it, you can easily open that last project by looking in File ➤ Recent Files and Solutions. You could also select File ➤ Open ➤ Project/Solution and browse to the project.

You may be wondering about the difference between a project and a solution. A solution is just a container that holds one or more projects. The projects can be of the same type, such as all SSRS projects. In some cases, the projects may be related to a topic area but may all be different technical types. For example, there could be a database project, an SSRS project, a SQL Server Integration Services Project, and a SQL Server Analysis Services project within the same solution. For example, the solution could be used to develop a data mart.

Depending on the value of a setting, the solution name will show up only if it contains more than one project. You can modify that setting if you choose by selecting Tools ➤ Options. In the Options dialog, expand Projects and Solutions and select the General page. If Always show solution is checked, the solution name will show up even if it contains one project. Figure 1-32 shows the dialog if you wish to make the settings change.

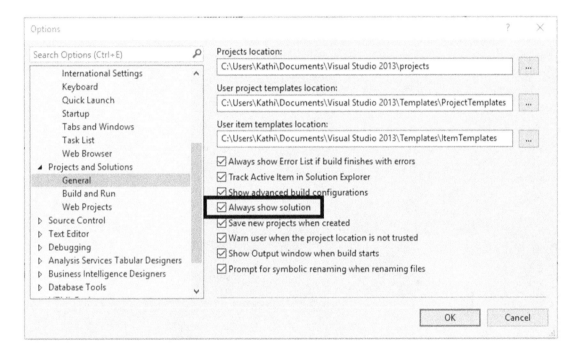

Figure 1-32. *The Always show solution option*

Summary

SQL Server Reporting Services is a wonderful feature of SQL Server that allows you to create reports that can be deployed for your organization's use. SSRS 2016 boasts a brand-new user interface called the web portal with the traditional paginated reports, key performance indicators (KPIs), and mobile reports.

To get your development environment set up, you will need to download and install several components. Luckily, they are all free downloads from Microsoft. By following the instructions in this chapter, you will be ready to learn how to develop and publish SSRS reports.

In Chapter 2, you will learn how to create your first reports by using a wizard.

CHAPTER 2

■ ■ ■

Using the Wizard to Create Your First Reports

Now that the environment is in place, you are ready to learn SQL Server Reporting Services (SSRS) report development. Before you begin to tackle the many details, you can take advantage of a built-in wizard that will enable you quickly create a report.

Software wizards ask the user a series of questions to automate a complex process. I like to compare them to the auto assembly line used by Henry Ford. Mr. Ford said, "Any customer can have a car painted any colour that he wants so long as it is black." You have many choices today when buying an automobile, but if you want something the manufacturer doesn't offer, you must add it later.

A wizard can do a lot of the work for you, but it is impossible to include everything that every report may require. Simple reports, with a bit of tweaking, may be good enough to deploy. At a minimum, it is a great way to start learning.

Creating Your First Report

Using the wizard, you can add new reports to existing projects. For this exercise, you will use it to create the project as well.

Follow these instructions to start the project:

1. Launch SQL Server Data Tools 2015. If you don't find it, and you have followed the instructions in Chapter 1, look for Visual Studio 2015 instead.

2. Click File ➤ New ➤ Project to launch the New Project dialog.

3. On the left, expand Installed, Templates, and Business Intelligence.

4. Select Reporting Services.

5. Select Report Server Project Wizard as shown in Figure 2-1.

© Kathi Kellenberger 2016

K. Kellenberger, *Beginning SQL Server Reporting Services*, DOI 10.1007/978-1-4842-1990-4_2

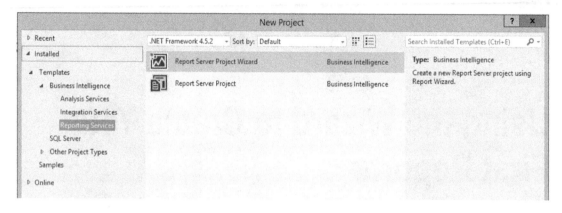

Figure 2-1. *Selecting the Report Server Project Wizard*

At the bottom of the dialog, you will fill in project and solution names and the location for the project. By default, the project is saved in a Projects folder in My Documents under Visual Studio 2015. I suggest that you create a destination specifically for working with this book in a location that you prefer but that will be easy for you to locate. For the Name, type in Wizard Reports. This is actually the name of the project. As you type in the project name, it will automatically fill in the Solution Name. Override that by typing in Beginning SSRS Chapter 2. Before clicking OK, make sure that the properties look similar to those in Figure 2-2. In this case, I have created a folder called Learn SSRS where all my projects and solutions related to this book will go.

Figure 2-2. *The name and location of the project*

After you click OK to create the project, the Report Wizard will launch. Follow these instructions to step through the wizard:

1. Click Next to move past the welcome screen.

2. The Select the Data Source screen allows you to set up connections to the data, called a data source. You will learn more about data sources in Chapter 3. For this exercise, click Edit shown in Figure 2-3 to bring up the Connection Properties dialog.

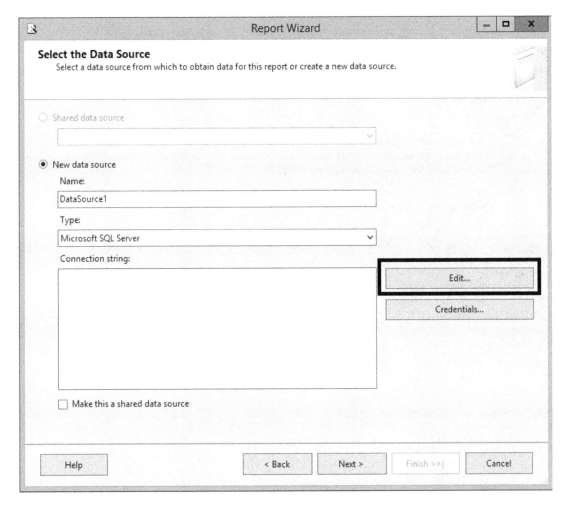

Figure 2-3. *The Select the Data Source dialog*

3. Make sure that the Data source property is set to Microsoft SQL Server (SqlClient).

4. Next you must fill in your Server name. You can specify a locally installed SQL Server by using (local), localhost, or a dot (.). If your instance has an instance name, such as Inst1, you will need to follow with a backslash and the instance name: (local)\Inst1. If you are not sure about the name of your instance, see the section "Determining the SQL Server Name" in Chapter 1. If your SQL Server instance is installed in your network, request help from your database administrator.

5. If your SQL Server is installed locally, accept the default of Use Windows Authentication in the Log on to the server property. Otherwise, check with your database administrator to see if you will use Windows authentication or if you will need to supply a user name and password.

6. In the Connect to a database section, choose Select or enter a database name and then find AdventureWorks2016 in the list.

■ **Note**   At the time of the SQL Server 2016 release, Microsoft had not made an AdventureWorks2016 database available. Instead, there was a beta version database called AdventureWork2016ctp3. If your database is named AdventureWorks2016CTP3, run this command in a new query window in SSMS to change the name:

```
ALTER DATABASE AdventureWorks2016CTP3 MODIFY NAME = AdventureWorks2016;
```

7.   Click Test Connection and click OK to dismiss the dialog if the test is successful. If not, you may need to ask your database administrator for help or make sure that you have the correct information supplied.

8.   The properties will look something like those in Figure 2-4. After reviewing, click OK to create the data source.

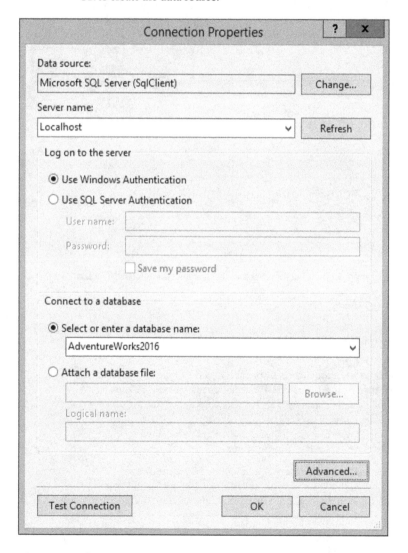

Figure 2-4. *The connection properties*

9. The Select the Data Source dialog will resemble Figure 2-5. The Connection String property will vary depending on your SQL Server instance.

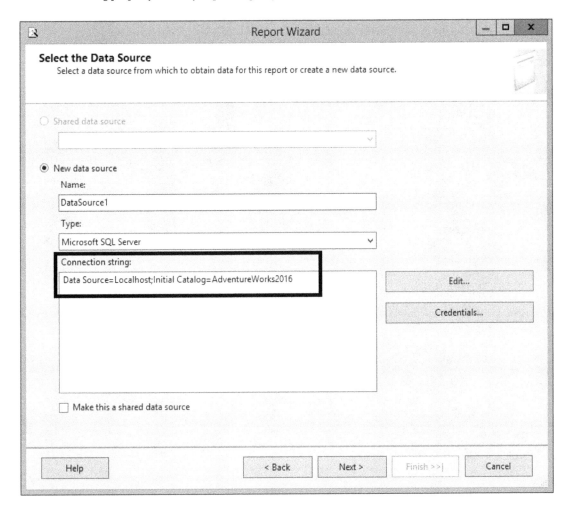

Figure 2-5. The Data Source property

10. Click Next to move to the Design the Query page. On this page, you can either use the Query Builder to visually create the query or you can directly type in a query with no extra assistance. Since this book is not meant to teach you T-SQL, the query has been provided. The book *Beginning T-SQL (Third Edition),* also written by the author (Apress, 2014), may be used to learn more about the T-SQL query language.

11. Type or paste in this code which is available in the Source Code/Download area of the Apress web site (Apress.com) and then click Next.

```
SELECT T.[Group], T.Name AS Region, YEAR(OrderDate) AS OrderYear,
    Month(OrderDate) AS OrderMonth, OrderDate, SalesOrderID, TotalDue
FROM Sales.SalesOrderHeader AS SOH
JOIN Sales.SalesTerritory AS T ON SOH.TerritoryID = T.TerritoryID;
```

12. On the Select the Report Type page, make sure that you select Tabular. You will create a Matrix report later in the chapter. Click Next.

13. On the Design the Table page, you will specify which pieces of information will make up the grouping levels of the report. You will learn more about grouping levels in Chapter 5. For now, configure the page to match Figure 2-6.

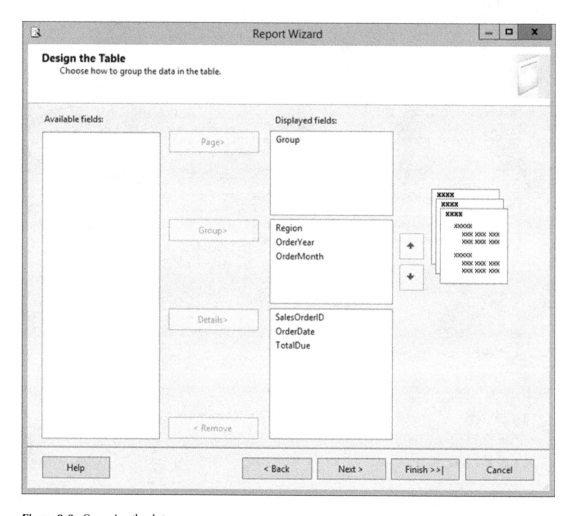

Figure 2-6. *Grouping the data*

14. Click Next to get to the Choose the Table Layout page. Make sure that Stepped, Include subtotals, and Enable drilldown are all selected as shown in Figure 2-7.

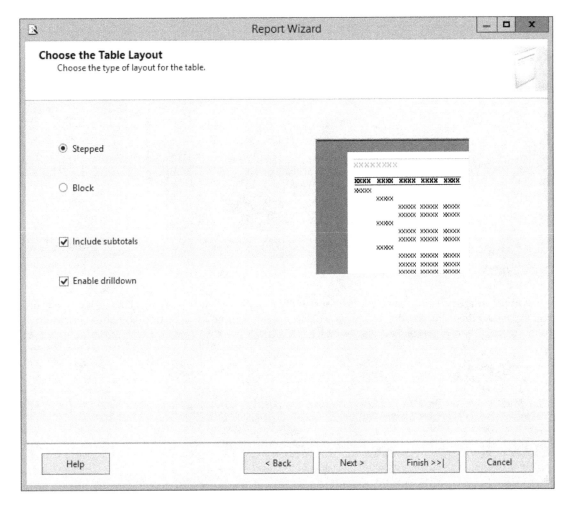

Figure 2-7. *The table layout*

15. On the final page of the wizard, type First Report in the Report Name property and click Finish.

▓ **Note** In previous versions of SQL Server, the wizard had an additional page to select a color scheme. At the time of SQL Server 2016 release, this option was not available.

Once the wizard is complete, you will have a new solution, project, and report. The report will be visible in design view as shown in Figure 2-8.

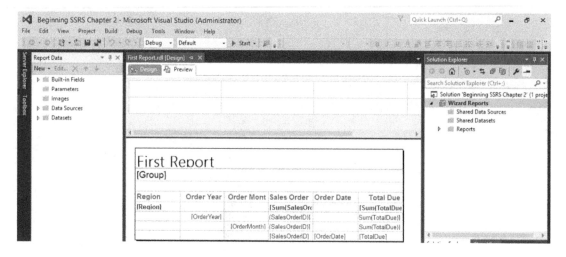

Figure 2-8. *The report in design view*

Within Visual Studio, you can view the report in three ways: design, preview, and code. To see the code, after expanding the Reports folder in Solution Explorer, right-click the name of the report and select View Code. Figure 2-9 shows part of the code file, which is XML.

```xml
<?xml version="1.0" encoding="utf-8"?>
<Report xmlns:rd="http://schemas.microsoft.com/SQLServer/reporting/reportdesigner" xmlns:c
  <AutoRefresh>0</AutoRefresh>
  <DataSources>
    <DataSource Name="DataSource1">
      <ConnectionProperties>
        <DataProvider>SQL</DataProvider>
        <ConnectString>Data Source=(local);Initial Catalog=AdventureWorks2016</ConnectStri
        <IntegratedSecurity>true</IntegratedSecurity>
      </ConnectionProperties>
      <rd:SecurityType>Integrated</rd:SecurityType>
      <rd:DataSourceID>9e9d9eae-6861-4a89-83c3-2cb8c5e86518</rd:DataSourceID>
    </DataSource>
  </DataSources>
  <DataSets>
    <DataSet Name="DataSet1">
      <Query>
        <DataSourceName>DataSource1</DataSourceName>
        <CommandText>SELECT T.[Group], T.Name As Region, YEAR(OrderDate) AS OrderYear, Mor
  OrderDate, SalesOrderID,  TotalDue
FROM Sales.SalesOrderHeader AS SOH
JOIN Sales.SalesTerritory AS T ON SOH.TerritoryID = T.TerritoryID;</CommandText>
        <rd:UseGenericDesigner>true</rd:UseGenericDesigner>
      </Query>
      <Fields>
```

Figure 2-9. *The report code*

Close the code window and click Preview to run the report. Figure 2-10 shows the first page of the report.

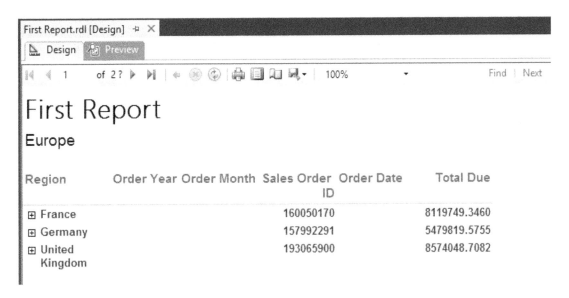

Figure 2-10. Page 1 of the report

Notice that you are looking at page 1, but the number of pages is two with a question mark. In order to return results more quickly to the end user, the beginning of the report may be returned before the middle and end are constructed. Click the right arrow to view additional pages. Once you reach the last page of the report, the question mark disappears, since now the page count is known, as shown in Figure 2-11.

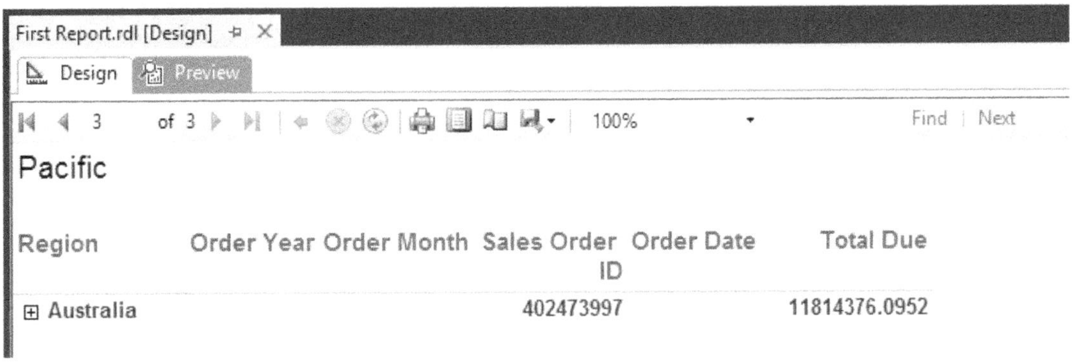

Figure 2-11. The last page of the report

On the Choose the Table Layout page of the wizard (see Figure 2-7), you checked the Enable drilldown property. This is why you see the plus sign next to Australia. Click the plus sign to expand the section. The Order Year data is now in view. Click to expand 2011 and Order Month 6. Now the details for that section are in view. To see how the drilldown property is set up, go to design view and select the Region row. In the Row Groups section beneath the report, right-click the table1_OrderYear group and select Group Properties. The Visibility page shows how the Region controls the visibility of the OrderYear group as shown in Figure 2-12. In practice, I have not found this to be a popular feature.

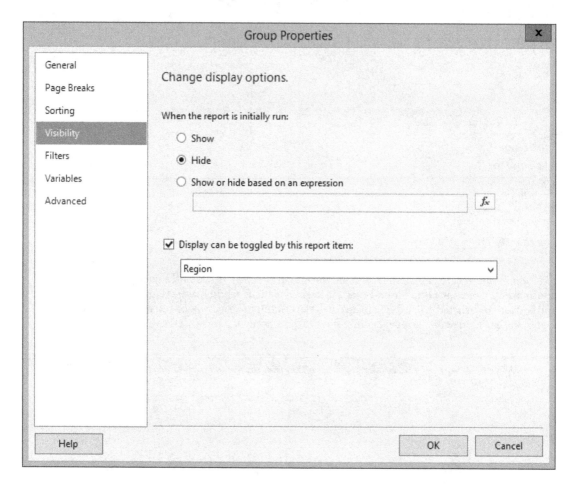

Figure 2-12. *The Visibility property*

Preview the report again and expand the sections. Figure 2-13 shows how the report should look.

Pacific

Region	Order Year	Order Month	Sales Order ID	Order Date	Total Due
⊟ Australia			402473997		11814376.0952
	⊟ 2011		20593854		1693032.7418
		⊞ 5	43701		3756.9890
		⊟ 6	2670115		227909.4738
			43703	6/1/2011 12:00:00 AM	3953.9884
			43704	6/1/2011 12:00:00 AM	3729.3640
			43705	6/1/2011 12:00:00 AM	3756.9890
			43709	6/2/2011 12:00:00 AM	3953.9884
			43710	6/2/2011 12:00:00 AM	3953.9884
			43715	6/4/2011 12:00:00 AM	3953.9884
			43716	6/4/2011 12:00:00 AM	3953.9884
			43717	6/4/2011 12:00:00 AM	772.5036
			43724	6/6/2011 12:00:00 AM	3953.9884

Figure 2-13. The report

Using the Preview Buttons

You may notice that there are some formatting issues with this report, but before you learn how to correct them, take a look at the buttons located above the report shown in Figure 2-14 that are available when you are previewing the report.

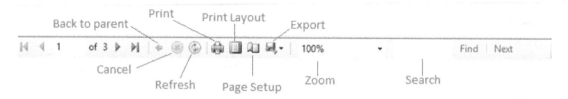

Figure 2-14. *The report preview icons*

To the right of the navigation controls, you will see an arrow pointing left, the Back to parent button. You will use this in Chapter 6 when you develop reports that link to each other. You will also see a button to stop execution of a long-running report and one to refresh the report. In the middle, there is a Print button. The most interesting buttons are found to the right of the Print button.

The Print Layout Button

After deploying a group of reports about a dozen years ago, I received a frantic phone call from the project manager. Why, she asked, was the report printing a blank page after every printed page? Of course, while previewing the reports from Visual Studio or Report Manager, the report looked perfect. It was only when printing did the problems surface.

In order to see just how the reports will look when printed, you should click the Print Layout button. This button, just to the right of the print button, toggles the view between the online view and the print view.

By scrolling through the report pages in print layout view, you will find problems that you can correct before the report is deployed. Make sure that you view every report you create in print layout before deploying it.

The Page Setup Button

The Page Setup button allows you to adjust the margins of the report and choose the paper size and source and the orientation. By default, the report will have one-inch margins. These margins may be too large. By modifying the properties, you will make sure that the report fits better on the printed page. You will learn about additional ways to fit reports on the page in Chapter 4. Figure 2-15 shows the Page Setup dialog box.

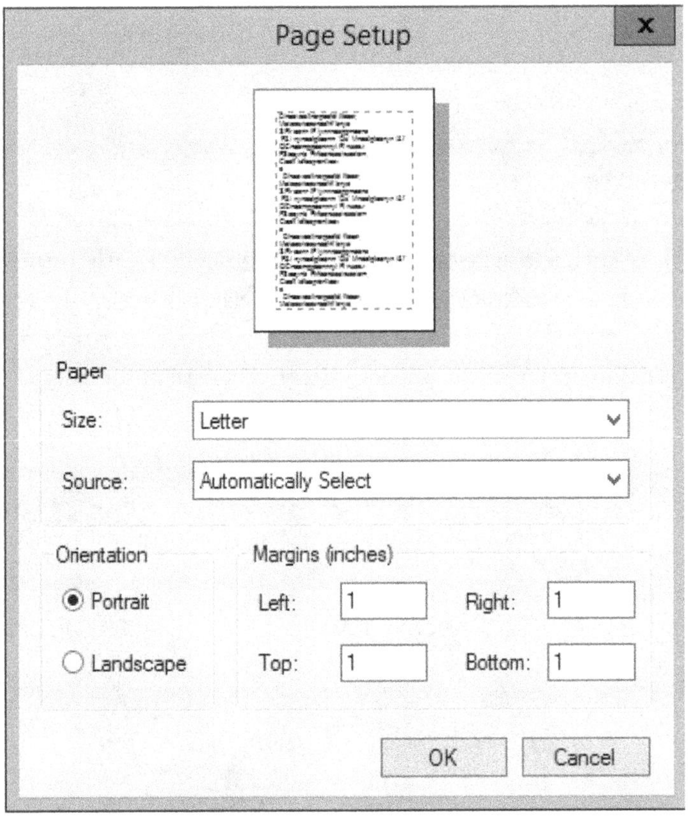

Figure 2-15. *The Page Setup dialog*

Additional Buttons

To the end user, exporting reports is just as important as printing them. An Export button with a variety of export types is available such as to XML, Excel, Word, and PDF. The results of the exported report will vary depending on the functionality of the export type.

You can magnify the preview window to several sizes with the Zoom button and search for text within the report with the Find and Next buttons.

Formatting the Wizard Report

The tabular report created with the wizard looks pretty good, but it can be improved. While looking at the preview, make a list of things to fix. This is what I came up with:

- Remove time from OrderDate
- Round dollar amounts to the nearest cent
- Format amounts as currency
- Remove summary values of SalesOrderID

- Change OrderMonth from numbers to month name

- Modify column widths

- Add a grand totals for each Group and the overall report.

Click Design to see the design view of the report. Find the OrderDate field in the bottom of the grid. Right-click the cell and select Text Box Properties.

■ **Note** It may take a bit of practice when selecting cells in the table grid. You can select the cell or the contents of the cell. When selecting the cell, it will be outlined. If you have trouble, try clicking the edge of the cell.

In the Text Box Properties dialog, select Number on the left. Under Category, choose Date. For the type, select a date format without the time as shown in Figure 2-16 and click OK.

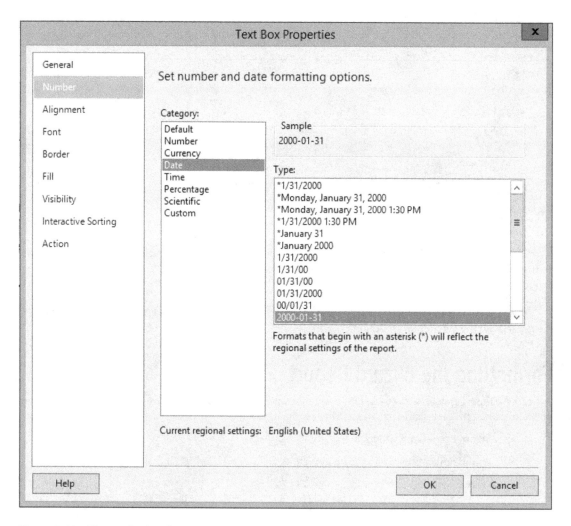

Figure 2-16. *Change the date format*

Right-click the TotalDue cell at the bottom right of the grid and select Text Box Properties. Select Number, but this time choose Currency. Check Use 1000 separator (,). The properties will look like those in Figure 2-17 if you are in the United States.

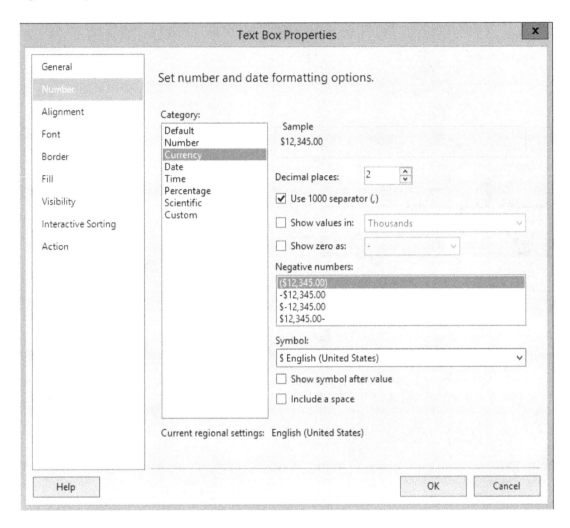

Figure 2-17. *Format the TotalDue field*

Click OK to accept the format. Now, to avoid repeating this process for the three summary fields, bring up the Properties window. You can bring this up by pressing F4 or find it in the View menu. Locate the Format property. Figure 2-18 shows what you are looking for.

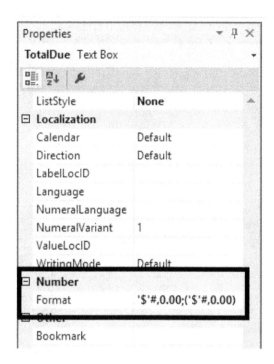

Figure 2-18. *The format property*

Copy the Format value to the Clipboard. Hold down the shift key and select the three summary boxes under the TotalDue heading. The property window is now selected for all three text boxes. Find the Format property and paste in the value.

Feel free to preview the report as often as you would like to check the progress. Each time you preview, the report definition is saved. At this point, the report should resemble Figure 2-19 after expanding the sections.

First Report

Europe

Region	Order Year	Order Month	Sales Order ID	Order Date	Total Due
⊟ France			160050170		$8119749.35
	⊟ 2011		3113568		$236268.63
		⊟ 5	43698		$3756.99
			43698	2011-05-31	$3756.99
		⊞ 6	525482		$40635.64
		⊞ 7	264026		$20120.82
		⊞ 8	530705		$40860.27
		⊞ 9	133211		$11861.97
		⊞ 10	580771		$44617.26

Figure 2-19. *The report with some formatting*

The wizard automatically created summaries for the SalesOrderID column since it contains numeric data. Summing up this value doesn't make sense. Go back to design view and remove [Sum(SalesOrderID)] by clicking within each cell with that formula and pressing the delete key. It needs to be removed from the second, third, and fourth rows.

The next item on the list is to change the Order Month column to display the month name. To change it to display the name instead, follow these steps:

1. In design view, right-click the OrderMonth cell and select Expression.

2. This brings up the Expression dialog box. You can see the existing expression, =Fields!OrderMonth.Value.

3. Expand Common Functions in the Category box.

4. Select Date & Time.

5. Position the cursor between the equal sign and the letter F in the expression.

6. Double-click MonthName in the Item list.

7. Add a closing parentheses at the end of the expression. The expression should now be =MonthName(Fields!OrderMonth.Value). The Expression dialog should look like Figure 2-20.

Figure 2-20. *The expression for month name*

8. Click OK after verifying the expression. Now that the expression is more complex, the cell value will change to <<Exp>>.

Some of the headings would look better if the columns were wider. When the table is selected, handles at the top and left appear. By clicking in the divisions between the columns of the header handle and dragging, you can expand or narrow the column widths to accommodate the column headings.

There is one more task on the list, adding a grand total and a total for each Group (US, Europe, and Pacific). This is easier to do than you might think. You'll learn more about working with grouping levels in Chapter 2, but understand at this point that a total added at one level lands at the next level up. For example, a total added at the Region level is displayed in the Group level. Follow these steps to add the Group totals:

1. In design view, right-click the cell located at Region and Total Due as shown in Figure 2-21.

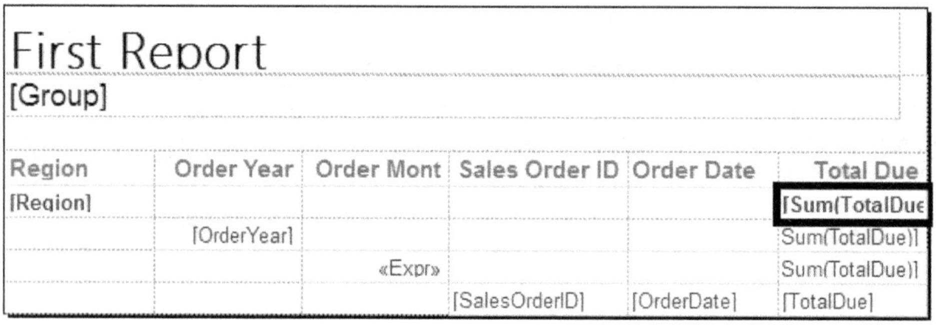

First Report
[Group]

Region	Order Year	Order Mont	Sales Order ID	Order Date	Total Due
[Region]					[Sum(TotalDue
	[OrderYear]				Sum(TotalDue)]
		«Expr»			Sum(TotalDue)]
			[SalesOrderID]	[OrderDate]	[TotalDue]

Figure 2-21. *The cell found at Region and Total Due*

2. Select Add Total. A new row will show up at the bottom of the grid containing the [Sum(TotalDue)] expression.

3. Type the following into the cell found at the bottom of the grid in the Sales Order ID column: Total for [Group]. Notice that you were able to include text and a field, called a placeholder, in the same cell without creating an expression.

The report definition should look like Figure 2-22.

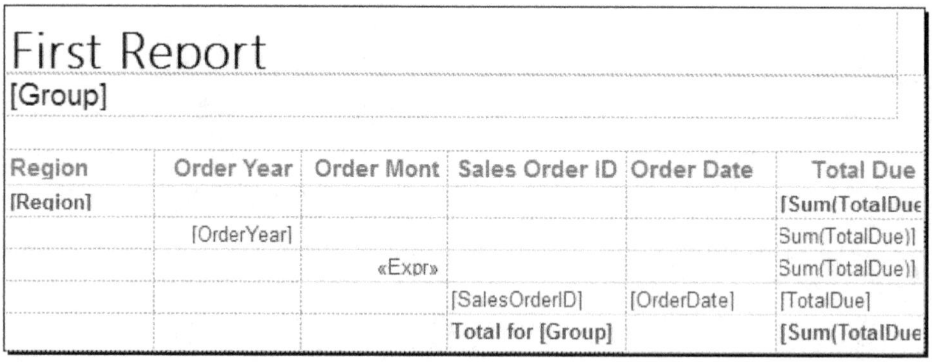

First Report
[Group]

Region	Order Year	Order Mont	Sales Order ID	Order Date	Total Due
[Region]					[Sum(TotalDue
	[OrderYear]				Sum(TotalDue)]
		«Expr»			Sum(TotalDue)]
			[SalesOrderID]	[OrderDate]	[TotalDue]
			Total for [Group]		[Sum(TotalDue

Figure 2-22. *The report definition after adding the Group total*

At the bottom of the design window, you will see the Row Groups and Column Groups sections, which you will learn more about in Chapter 5. Grouping levels can be added and configured in these sections. Select the Group text box at the top of the report which changes what appears in the Row Groups section. Right-click list1_Group under Row Groups, and select Add Total ➤ After as shown in Figure 2-23.

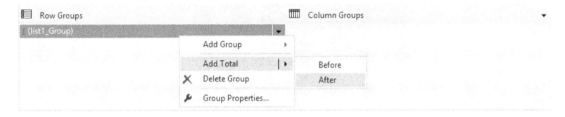

Figure 2-23. *Adding a grand total*

This just adds a row where you will need to add the expression. Right-click the new text box at the bottom of the report and select Expression. Type in this expression and click OK:

```
="Grand Total " & FormatCurrency(Sum(Fields!TotalDue.Value),2,0,True,True)
```

At the top of the screen, you will see a formatting menu bar as shown in Figure 2-24. This menu bar is similar to working in other programs such as Microsoft Word.

Figure 2-24. *The formatting menu bar*

Make sure that the new text box is selected and click the icon to align right. Click the B icon to bold the text. The report definition should now resemble Figure 2-25.

First Report
[Group]

Region	Order Year	Order Mont	Sales Order ID	Order Date	Total Due
[Region]					[Sum(TotalDue
	[OrderYear]				Sum(TotalDue)]
		«Expr»			Sum(TotalDue)]
			[SalesOrderID]	[OrderDate]	[TotalDue]
			Total for [Group]		[Sum(TotalDue
					«**Expr**»

Figure 2-25. *The report definition after adding totals*

Now preview the report. Each page will have a total for the group. The very last page will have a grand total. Figure 2-26 shows page 3 of the report with the new totals.

Pacific					
Region	Order Year	Order Month	Sales Order ID	Order Date	Total Due
⊞ Australia					$11,814,376.10
			Total for Pacific		$11,814,376.10
				Grand Total	$123,216,786.12

Figure 2-26. *Page 3 of the formatted report with the new totals*

Creating a Matrix Report

During the previous exercise, you created the solution and project along with the report while running the wizard. You can also add a new report to an existing project with the wizard. In this section, you will add a matrix report to the project.

A matrix report is often a more compact report than a tabular report. In a matrix report, one or more of the columns of the data will be pivoted to become a column header. In this example, the data will be pivoted by year. To show how this works, view the results of this query that returns the total sales by year for North America.

```
SELECT T.[Group], T.Name As Region, YEAR(OrderDate) AS OrderYear,
    SUM(TotalDue) AS TotalSales
FROM Sales.SalesOrderHeader AS SOH
JOIN Sales.SalesTerritory AS T ON SOH.TerritoryID = T.TerritoryID
WHERE T.[Group] = 'North America'
GROUP BY T.[Group], T.Name , YEAR(OrderDate)
ORDER BY Region, OrderYear;
```

You can see the partial results of the query in Figure 2-27.

	Group	Region	OrderYear	TotalSales
1	North America	Canada	2011	2106905.8728
2	North America	Canada	2012	6599971.0217
3	North America	Canada	2013	7010449.6994
4	North America	Canada	2014	2681602.5941
5	North America	Central	2011	1126645.7497
6	North America	Central	2012	3334867.9788
7	North America	Central	2013	3374336.2992
8	North America	Central	2014	1077449.2196

Figure 2-27. *The partial results of the total sales by year*

In order to compare by year—in other words, to easily compare the 2011 Canadian sales to the 2011 Central sales—you can pivot the data. Here is a query that pivots the values of OrderYear into columns.

```
SELECT [Group], Region,
    [2011], [2012], [2013], [2014]
FROM
(SELECT T.[Group], T.Name As Region, YEAR(OrderDate) AS OrderYear,
    SUM(TotalDue) AS TotalSales
FROM Sales.SalesOrderHeader AS SOH
JOIN Sales.SalesTerritory AS T ON SOH.TerritoryID = T.TerritoryID
WHERE T.[Group] = 'North America'
GROUP BY T.[Group], T.Name , YEAR(OrderDate)
) AS SourceTable
PIVOT
(SUM(TotalSales)
FOR OrderYear IN ([2011], [2012], [2013], [2014])
)
AS PivotTable
ORDER BY Region;
```

Figure 2-28 shows the results.

	Group	Region	2011	2012	2013	2014
1	North America	Canada	2106905.8728	6599971.0217	7010449.6994	2681602.5941
2	North America	Central	1126645.7497	3334867.9788	3374336.2992	1077449.2196
3	North America	Northeast	705672.1952	3272239.7992	2965567.0284	876730.6057
4	North America	Northwest	2620943.826	5325813.0562	6759500.6713	3355402.8175
5	North America	Southeast	1847744.578	3344683.6085	2705730.9695	985940.2109
6	North America	Southwest	3144713.0989	9329154.3425	10239209.3403	4437517.8076

Figure 2-28. *The pivoted results*

The T-SQL pivot query syntax is complex, and you must hard-code the column names in the query. Fortunately, creating pivoted results in SSRS is really easy, and no hard-coding is involved.

Follow these steps to create the matrix report:

1. To kick off the wizard within the project, right-click the Reports folder in the Solution Explorer window and select Add New Report as shown in Figure 2-29.

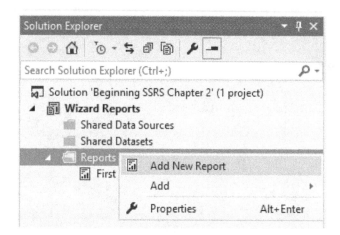

Figure 2-29. How to kick off the wizard

2. Once the wizard starts up, click Next to go past the welcome page.

3. On the Select the Data Source page, click Edit to bring up the Connection Properties. Fill in the connection information as was done in the section "Creating Your First Report" earlier in this chapter.

4. After configuring the data source, click Next. In the Design the Query page, use the same query that was used in the earlier example:

```
SELECT T.[Group], T.Name As Region, YEAR(OrderDate) AS OrderYear,
    Month(OrderDate) AS OrderMonth,
    OrderDate, SalesOrderID, TotalDue
FROM Sales.SalesOrderHeader AS SOH
JOIN Sales.SalesTerritory AS T ON SOH.TerritoryID = T.TerritoryID;
```

5. Click Next to move to the Select the Report Type page. Select Matrix and click Next.

6. Configure the Design the Matrix page as shown in Figure 2-30. When you select a field on the right, the section of the report where it will show is highlighted. This can help you figure out which fields go where.

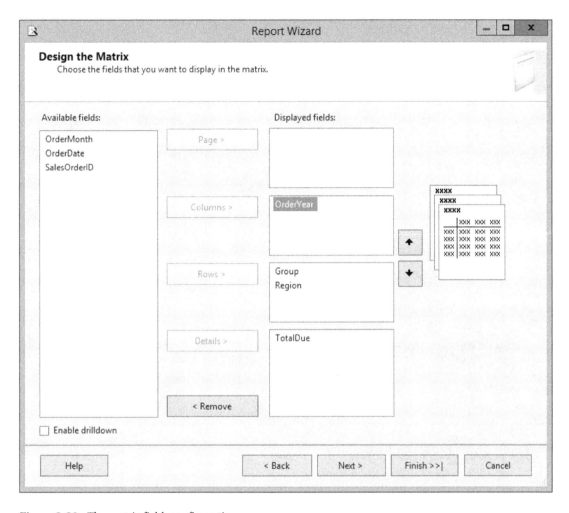

Figure 2-30. *The matrix fields configuration*

7. Click Next and fill in a report name. Call this Matrix Report.

8. Click Finish to complete the wizard.

Now you will see two reports in the Reports folder of the Solution Explorer. To open a report, double-click the name. You can also switch between open reports in the design area. To see how the matrix report turned out, click Preview. The report should look like Figure 2-31.

Matrix Report

		2011	2012	2013	2014
Europe	France	236268.6270	1743487.6538	4271019.2663	1868973.7989
	Germany	272780.9066	607828.1753	2869491.9712	1729718.5224
	United Kingdom	400991.9290	1769769.2149	4068178.6672	2335108.8971
North America	Canada	2106905.8728	6599971.0217	7010449.6994	2681602.5941
	Central	1126645.7497	3334867.9788	3374336.2992	1077449.2196
	Northeast	705672.1952	3272239.7992	2965567.0284	876730.6057
	Northwest	2620943.8260	5325813.0562	6759500.6713	3355402.8175
	Southeast	1847744.5780	3344683.6085	2705730.9695	985940.2109
	Southwest	3144713.0989	9329154.3425	10239209.3403	4437517.8076
Pacific	Australia	1693032.7418	2347885.4611	4702404.0504	3071053.8419

Figure 2-31. *View the matrix report*

The big difference between tabular and matrix reports is that matrix reports have grouping levels across the columns. You will learn much more about matrix reports in Chapter 5. In the meantime, go back to the design view of the report to view the properties. Figure 2-32 shows how simple the design of the matrix report really is.

Matrix Report

	[OrderYear]	
[Group]	[Region]	[Sum(TotalDue)]

Figure 2-32. *The matrix report design*

Using the skills you have learned earlier in the chapter, format the summary textbox. To make this report more complete, you will add three total fields. Follow these instructions to add the totals.

1. Right click on the textbox containing the expression [Sum(TotalDue)].

2. Choose Add Total ➤ Row.

3. Now repeat the process but this time select Add Total ➤ Column.

4. Right-click the cell at the intersection of Region and Total. Select Add total.

5. Select the Total column and click B in the design menu to bold the font.

6. Select the bottom row and click B to bold the font.

7. Right align the first row.

The report design will now look like Figure 2-33.

Matrix Report

	[OrderYear]	**Total**
[Group]	[Region] [Sum(TotalDue)	[Sum(TotalDue
	Total [Sum(TotalDue	[Sum(TotalDue

Figure 2-33. *The report design after adding the totals*

Preview the report. You now have totals for all the regions and totals across the years. Figure 2-34 shows the report.

Matrix Report

		2011	2012	2013	2014	**Total**
Europe	France	$236,268.63	$1,743,487.65	$4,271,019.27	$1,868,973.80	**$8,119,749.35**
	Germany	$272,780.91	$607,828.18	$2,869,491.97	$1,729,718.52	**$5,479,819.58**
	United Kingdom	$400,991.93	$1,769,769.21	$4,068,178.67	$2,335,108.90	**$8,574,048.71**
	Total	**$910,041.46**	**$4,121,085.04**	**$11,208,689.90**	**$5,933,801.22**	**$22,173,617.63**
North America	Canada	$2,106,905.87	$6,599,971.02	$7,010,449.70	$2,681,602.59	**$18,398,929.19**
	Central	$1,126,645.75	$3,334,867.98	$3,374,336.30	$1,077,449.22	**$8,913,299.25**
	Northeast	$705,672.20	$3,272,239.80	$2,965,567.03	$876,730.61	**$7,820,209.63**
	Northwest	$2,620,943.83	$5,325,813.06	$6,759,500.67	$3,355,402.82	**$18,061,660.37**
	Southeast	$1,847,744.58	$3,344,683.61	$2,705,730.97	$985,940.21	**$8,884,099.37**
	Southwest	$3,144,713.10	$9,329,154.34	$10,239,209.34	$4,437,517.81	**$27,150,594.59**
	Total	**$11,552,625.32**	**$31,206,729.81**	**$33,054,794.01**	**$13,414,643.26**	**$89,228,792.39**
Pacific	Australia	$1,693,032.74	$2,347,885.46	$4,702,404.05	$3,071,053.84	**$11,814,376.10**
	Total	**$1,693,032.74**	**$2,347,885.46**	**$4,702,404.05**	**$3,071,053.84**	**$11,814,376.10**

Figure 2-34. *The matrix report with totals*

Summary

Most reports, especially if they take advantage of some of the advanced features of SSRS, cannot be created with the wizard. The wizard is, however, perfect for simple reports. It is also a great tool to help you begin learning to develop SSRS reports. In this chapter, you created two reports. You then modified the properties so that they looked professional and were ready to be deployed.

In Chapter 3, you will learn how to configure data sources and datasets, the most basic elements of reports.

PART II

Report Development

CHAPTER 3

Understanding Data Sources and Datasets

I deployed my first group of SSRS reports back in 2004 when SQL Server Reporting Services (SSRS) was initially released. The reports were so successful that word got around to the other departments at the company where I worked. Soon I had more requests for reports than I could handle. I was a database administrator first, and creating reports was just a small part of my job. It wasn't long before I put together a day-long SSRS workshop to teach a small group of people from each department how to develop their own reports.

Since then, I have taught SSRS development to dozens of people with classes, articles, and my original SSRS book. Knowledge of data sources and datasets, covered in this chapter, is fundamental to building reports. When teaching SSRS, I learned that these topics are challenging to understand as well. Make sure you fully comprehend the material covered in this chapter before moving on, and be sure to refer back whenever you need help.

This chapter covers both data sources and datasets. You will learn what they are, how to create them, and when it makes sense to share them among reports. In Chapter 2, you created the project and two reports by running the wizard. In this chapter, you will build the reports manually.

Creating Shared Data Sources

If you have worked with databases before, you are probably familiar with the concept of a connection string. A connection string is like an address telling you everything you need to communicate with a database, such as the type of database, the location, and the account used to connect. Here is a possible connection string:

```
Provider=SQLNCLI11;Server=localhost\INST1;Database=AdventureWorks2016;
Trusted_Connection=yes;
```

This connection says that the type of connection is for SQL Server using the latest version of the Native Client. The server is located on the local computer and it is a named instance, INST1. The database to connect to is AdventureWorks2016. Instead of providing a user name and password, Windows credentials will be used.

The connection string may look intimidating, but you don't have to figure it out yourself. Most applications, SSRS included, have an intuitive tool for building connection strings.

In Chapter 2, you created a solution with one project. Inside the project, you created two reports. To keep things simple and to focus on creating a report with the wizard, I had you embed each data source within a report. This means that the data source became part of the report definition. If you look in the XML source code of those reports, you will find the connection string.

© Kathi Kellenberger 2016
K. Kellenberger, *Beginning SQL Server Reporting Services*, DOI 10.1007/978-1-4842-1990-4_3

Most of the time, data sources should be shared among reports instead. Imagine that you have 100 or more reports with the identical embedded connection string in each one. Then you get word that the database is moving to a new server over the weekend. You are looking at hours of work editing and testing each report. If the data source was shared instead of embedded, you would need to make just one simple change, and all of your reports would be pointing to the new database location.

As a best practice, data sources should be shared, and SSRS makes it easy to do so. To learn how to create a shared data source for your project, take the following steps:

1. Launch SQL Server Data Tools 2015 (SSDT).

2. Create a new project by clicking File ➤ New Project.

3. Once the New Project dialog opens, find the Reporting Services templates under Installed ➤ Templates ➤ Business Intelligence.

4. Select Report Server Project as shown in Figure 3-1.

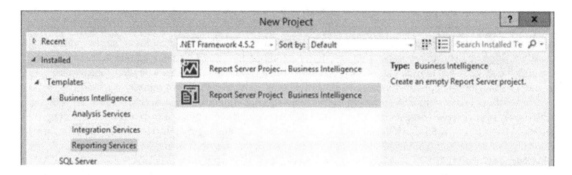

Figure 3-1. The Report Server Project

5. Fill in Data Sources and Datasets for the project name.

6. Fill in Beginning SSRS Chapter 3 for the solution name as shown in Figure 3-2.

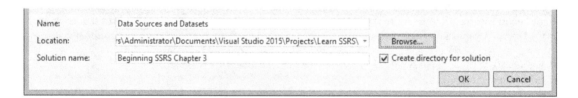

Figure 3-2. The project and solutions names

7. Make sure that the solution will be saved where you can find it. It will be created in the location of the previous solution by default. Click OK to create the project.

8. Make sure that you can see the Solution Explorer window. If it is not visible, enable it from the View menu. You can also use the keyboard shortcut CTRL + ALT + L. Figure 3-3 shows how the Solution Explorer should look.

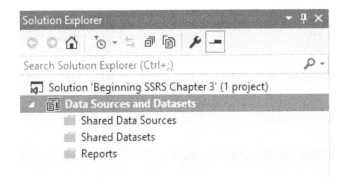

Figure 3-3. *The Solution Explorer*

The project can contain three types of objects: Shared Data Sources, Shared Datasets, and Reports. The only real requirement is to have a report; the data sources and datasets may be embedded in the reports. In that case, you will not see them in these folders. To add a Shared Data Source, follow these steps:

1. Right-click Shared Data Sources and select Add New Data Source as shown in Figure 3-4.

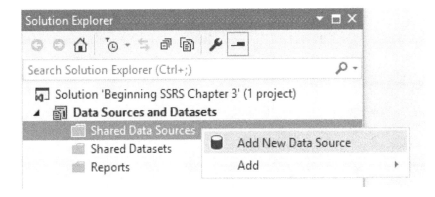

Figure 3-4. *Add a new data source*

2. This brings up the Shared Data Source Properties dialog as shown in Figure 3-5. Fill in Name: for the data source, AdventureWorks2016. Naming each data source is very important so that the data source can be identified later.

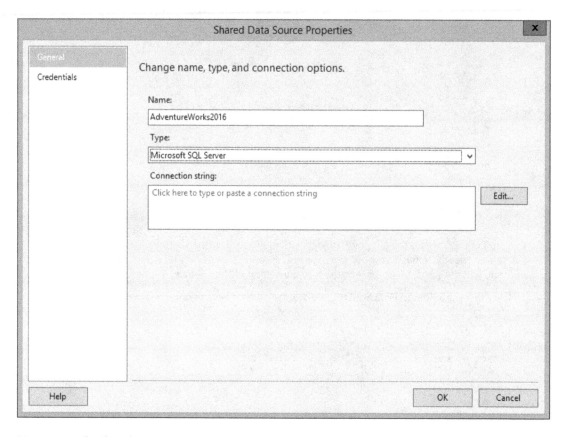

Figure 3-5. *The Shared Data Source Properties*

3. The Type of data source should be Microsoft SQL Server.

4. Click the Edit button to bring up the Connection Properties dialog.

5. Fill in the server name and logon properties for your SQL Server instance. If you have trouble figuring out your server instance name, see the section "Determining the SQL Server Name," in Chapter 1 or ask your database administrator for assistance.

6. Under Select or enter a database name, select the AdventureWorks2016 database. The properties should resemble Figure 3-6.

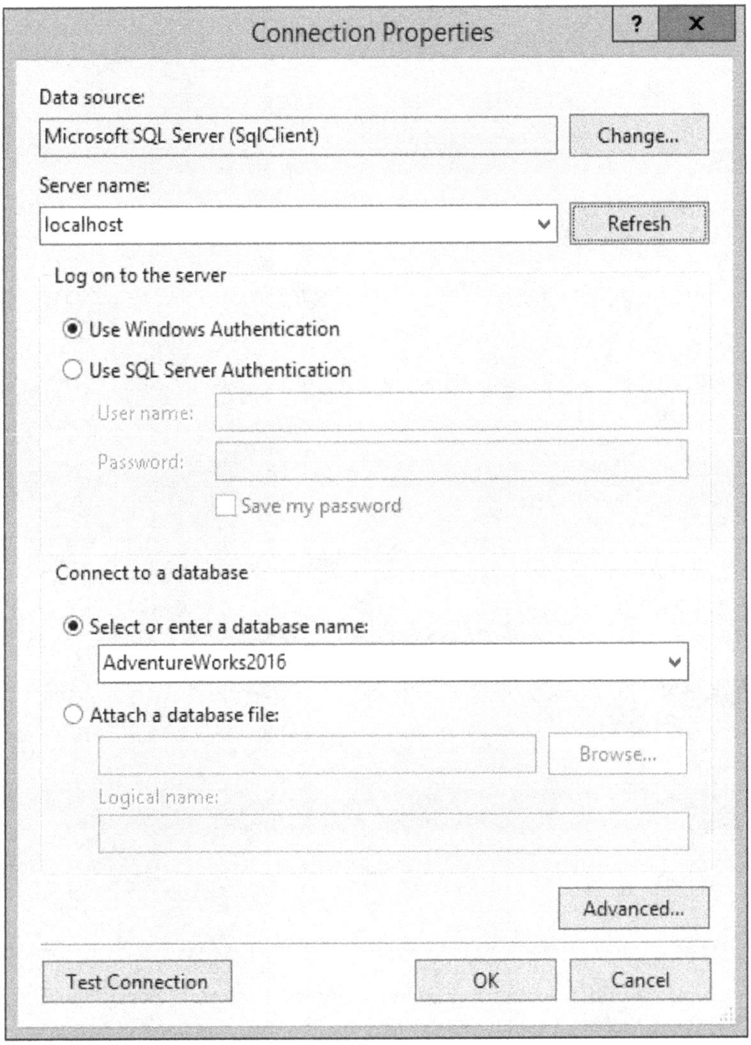

Figure 3-6. *The connection properties*

■ **Note** Going forward, the examples in the book will assume that you know how to connect to your database.

---

7. Click Test Connection to make sure that the connection information is correct. Click OK to dismiss the test box and OK again to accept the connection properties. The Connection string property should now be filled in.

Click Ok to accept the data source's properties. The data source will now be visible in the Solution Explorer as shown in Figure 3-7.

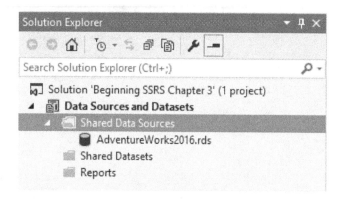

Figure 3-7. *The new data source*

Notice that the data source is actually an XML file. You don't have the option to view the file's contents in Visual Studio, but you can navigate to the rds file and open it up in a text editor if you wish. Figure 3-8 shows the contents of my data source file.

```xml
<?xml version="1.0" encoding="utf-8"?>
<RptDataSource xmlns:xsi="http://www.w3.org/2001/XMLSchema-instance" xmlns:xsd="h
  <ConnectionProperties>
    <Extension>SQL</Extension>
    <ConnectString>Data Source=localhost;Initial Catalog=AdventureWorks2016</Conr
    <IntegratedSecurity>true</IntegratedSecurity>
  </ConnectionProperties>
  <DataSourceID>fc5ecc95-b9e4-4394-8c92-2b7ef713207a</DataSourceID>
</RptDataSource>
```

Figure 3-8. *The data source file*

If you need to edit the data source properties, click the data source or right-click it and select Open. It may be a bit counterintuitive, but if you right-click and select Properties, you will see only the File Name and File Path in the Properties window as shown in Figure 3-9.

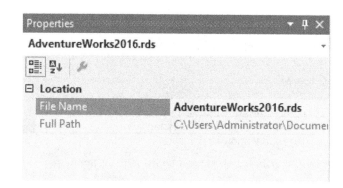

Figure 3-9. *The properties of the data source*

Set up any connections that will be needed by the reports in your project. As a best practice, name the shared data sources so that they can be identified later. You should avoid naming the data sources with the server name. Often you will be doing development against a local, development, or testing server. The published reports will usually point to a production server with a different name.

■ **Note** If you would like to use the projects found in the code download, you may need to modify the data source properties in order to connect to your own SQL Server instance.

Creating Shared Datasets

A dataset is a query, the question you are asking the database. Unlike data sources, datasets are usually unique to the reports and most should not be shared. There are a couple of exceptions, however. For example, you may need to reuse a parameter list in multiple reports. There are also some new features, such as mobile reports, that required shared datasets. In this example, you will create a parameter list as a shared dataset.

To create a shared dataset follow these steps:

1. Right-click the Shared Datasets folder and select Add New Dataset.

2. This brings up the Shared Dataset Properties dialog. Type Year in the Name property.

3. Make sure that the data source you created in the last section, AdventureWorks2016, is selected in the Data source property.

4. Make sure that Text is chosen as the Query type property.

5. Type the following code in the Query text box:

```
SELECT DISTINCT YEAR(OrderDate) AS OrderYear
FROM Sales.SalesOrderHeader
ORDER BY OrderYear;
```

6. The properties should look like Figure 3-10. Click OK to create the new shared dataset.

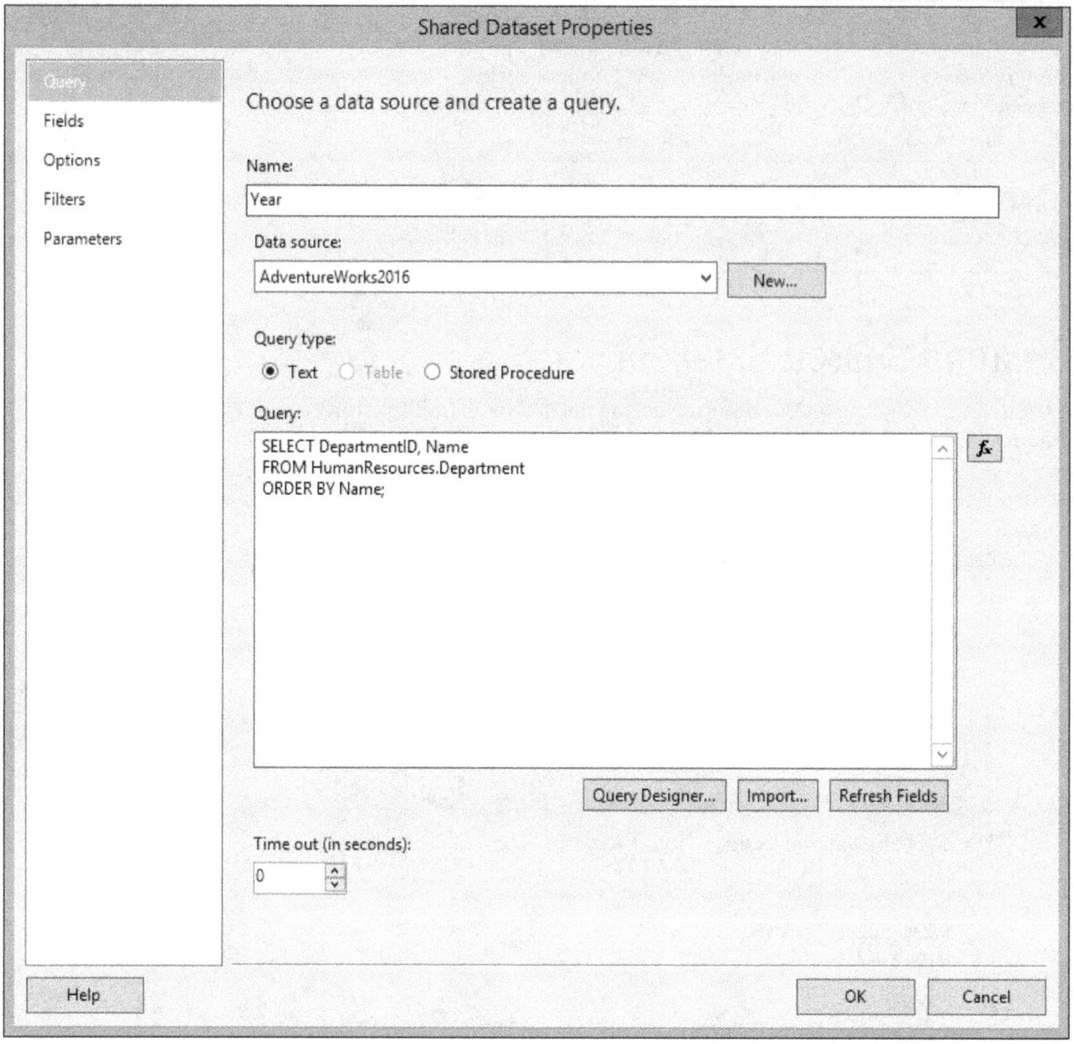

Figure 3-10. *The Shared Dataset Properties*

You should now see the Year dataset in the Shared Datasets folder as shown in Figure 3-11. Just like the data sources, viewing the properties shows the file location. To edit the properties, click the name or right-click and select Open.

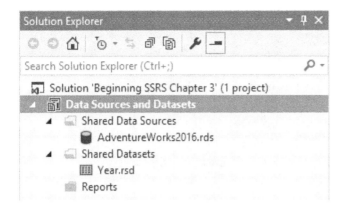

Figure 3-11. *The new shared dataset*

You can create additional datasets as required. Remember that it doesn't make sense to share a dataset unless it will be used in multiple reports. The typical use case is for a parameter list.

■ **Note**   At the release of SQL Server 2016, there was a bug involving Shared Datasets. To fix the problem, navigate to the project files and open the Year.rsd file. Change this code `<Dataset>` to `<Dataset Name="Year">` and then save the file. Microsoft has promised to fix this issue soon, so it may not be a problem by the time you are reading this book.

Using Data Sources and Datasets

The reason for creating data sources and datasets is to use them in reports. Now you will do just that. Follow these steps to create the report:

1. Right-click the Reports folder and select Add ➤ New Item as shown in Figure 3-12. Make sure that you do not choose Add New Report which kicks off the wizard.

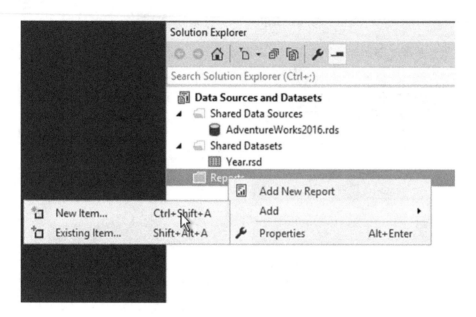

Figure 3-12. *Add a new report*

2. In the Add New Item – Data Sources and Datasets dialog, select Report.

3. Fill in the report name property with Sales by Year. The dialog should look like Figure 3-13.

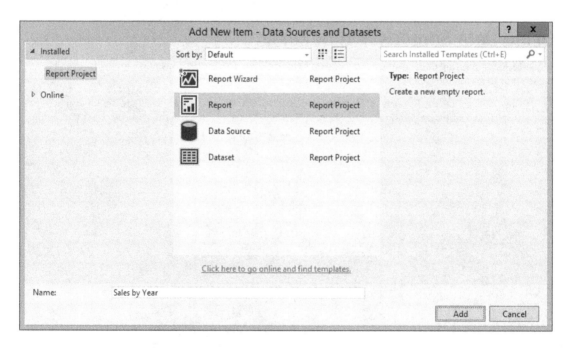

Figure 3-13. *The new report properties*

4. Click Add to create the new report.

The report should now be visible in the Solution Explorer and open in design view. If not, double-click the report name to open it. The next task is to set up the data source within the report so that it points to the shared data source.

Find the Report Data window which will probably be located on the left. If it is not visible, click the design canvas of the report, and then click Report Data found at the bottom of the View menu. The Report Data window has several folders as shown in Figure 3-14.

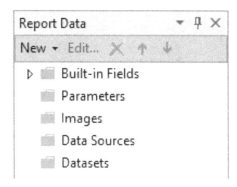

Figure 3-14. *The Report Data window*

Follow these steps to set up the report's data source:

1. Right-click the Data Sources folder and select Add Data Source.

2. This brings up the Data Source Properties dialog box. Fill in the Name property with AdventureWorks. As a best practice, always give each data source a descriptive name. This data source will be linked to the AdventureWorks2016 shared data source. You can, if you wish, give it the exact same name. In this case, you will change the name slightly so it is easy to see the difference between the shared data source and the data source in the report.

3. The middle of the dialog lets you create this report data source as an embedded data source with connection properties visible to only this report. Instead, select Use data source reference. Find AdventureWorks2016 in the list. The dialog should look like Figure 3-15.

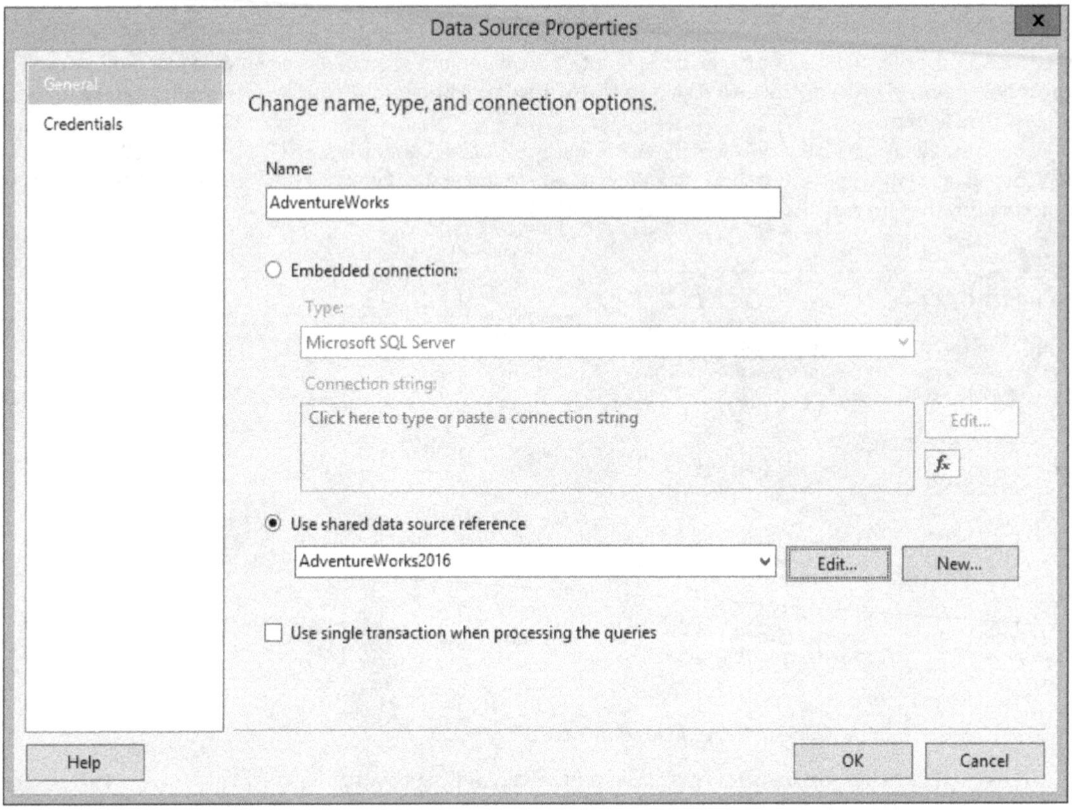

Figure 3-15. *The report data source properties*

4. Click OK to create the data source.

The Edit button on the dialog shown in Figure 3-15 allows you to modify the shared data source if you need to. Remember that changes made this way will affect all reports using this data source. You can also create a new shared data source. As a best practice, and to avoid confusion, always create and edit shared data sources at the project level.

You should now see the AdventureWorks data source in the Data Sources folder in the Report Data window as shown in Figure 3-16. The tiny arrow on the icon signifies that it is pointing to a shared data source.

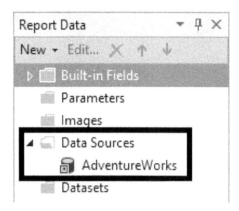

Figure 3-16. *The new data source*

The next step in creating a report is to add a dataset. Datasets are queries, the question to the database or other source of data. A dataset will usually be specific to a report, so it doesn't make sense to share most of them. To create the first dataset for this report, follow these steps:

1. Right-click the Datasets folder and select Add Dataset. This brings up the Dataset Properties dialog box. By default it is set to Use a Shared Dataset as shown in Figure 3-17.

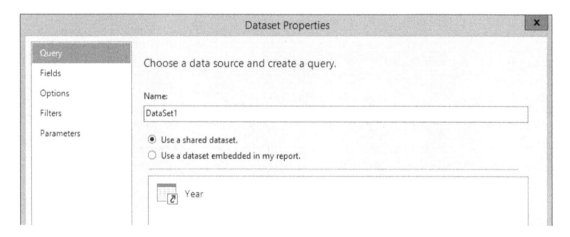

Figure 3-17. *The Dataset Properties dialog*

2. Fill in the Name property with Sales.

3. Select Use a Dataset embedded in my report. When you do, the dialog changes. Instead of showing the shared dataset, the dialog displays the properties for the embedded dataset. Figure 3-18 shows how the dialog looks after the changes.

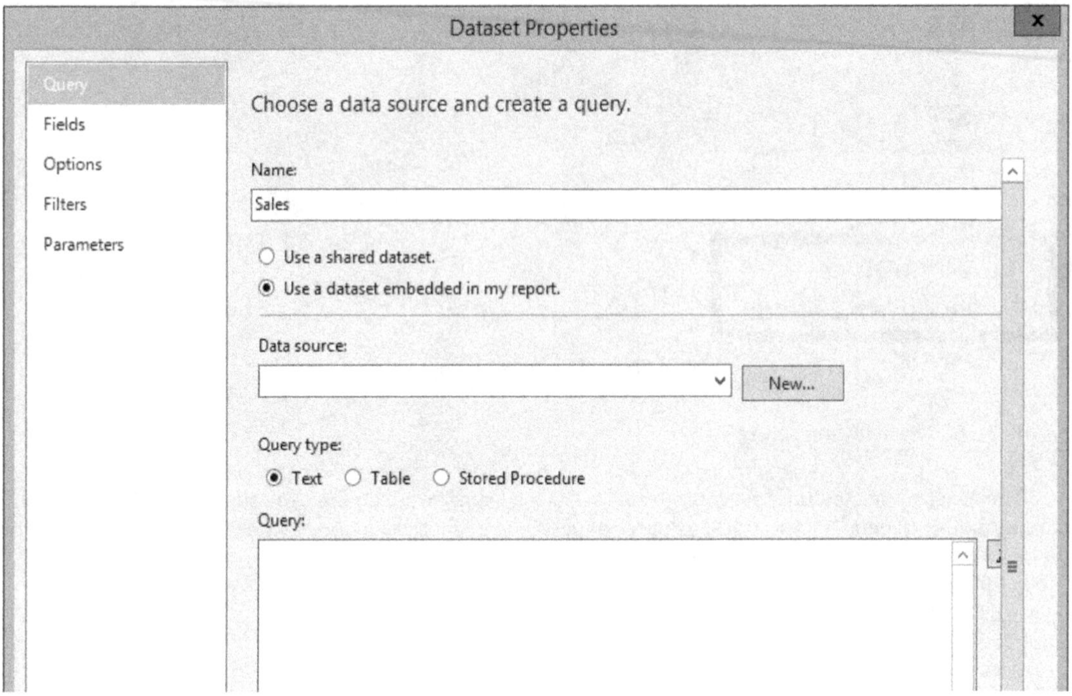

Figure 3-18. *The Dataset Properties for an embedded dataset*

4. Find AdventureWorks in the dropdown box under Data source. Only data sources that have been set up as part of the report are visible in this list.

5. Select Text as the Query type if it is not selected.

6. Type this code in the Query property:

```
SELECT SUM(TotalDue) AS TotalSales, MONTH(OrderDate) AS OrderMonth,
YEAR(OrderDate) AS OrderYear
FROM Sales.SalesOrderHeader
GROUP BY MONTH(OrderDate), YEAR(OrderDate);
```

7. Click OK to create the dataset.

The Sales dataset should now be visible in the Report Data window as shown in Figure 3-19. In addition to the dataset name, you will also see the available fields.

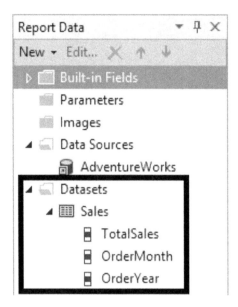

Figure 3-19. *The Sales dataset*

Now you can display data from the dataset on the report. Follow these steps to set up a simple report:

1. Display the Toolbox window. It if is not visible, click the report design canvas and then click Toolbox from the View menu. The Toolbox looks like Figure 3-20.

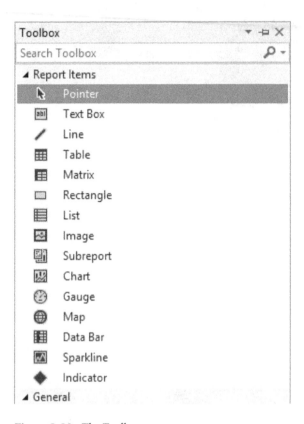

Figure 3-20. The Toolbox

2. Drag a Table control from the Toolbox on to the report design canvas. You can also right-click the design surface and add a table by selecting Insert ➤ Table. Figure 3-21 shows the table control on the report.

Figure 3-21. The Table control

3. There are several ways to populate the table. From the Report Data window, drag the OrderYear field to the leftmost cell in the Data row. The Header row automatically populates.

4. Hover over the middle cell in the Data row until a small icon appears in the cell. Click the icon to display a list of the available fields. Select OrderMonth.

5. Click in the cell on the right in the Data row. Type in this code:

    [TotalSales]

6. In the header row above, type Total Sales. The table grid should look like Figure 3-22.

Order Year	Order Month	Total Sales
[OrderYear]	[OrderMonth]	[TotalSales]

Figure 3-22. *The table with cells populated*

Click Preview to view the report. Obviously, this report is not ready to be published, but it does demonstrate the steps required to hook a report to the data.

Using a Shared Dataset

Shared datasets are useful for queries that will be reused throughout the project. SSRS allows you to create reports that let the user running the report to filter dynamically. Often, the same criteria will be used with multiple reports which is a great use of shared datasets. Follow these steps to add the shared dataset:

1. Switch back to Design view.

2. In the Report Data window, right-click Datasets and select Add Dataset.

3. Name the dataset Year.

4. Make sure that Use a shared dataset is chosen and select Year from the window. The Dataset Properties dialog should look like Figure 3-23.

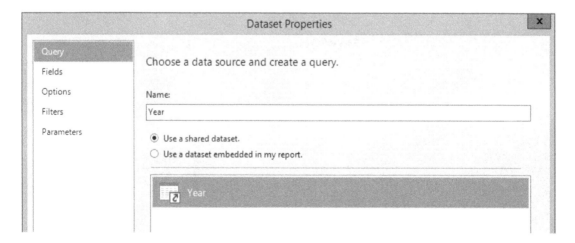

Figure 3-23. *Using a shared dataset*

5. Click OK to create the new dataset.

The next step is to alter the Sales dataset so that it requires a parameter. Double-click the Sales dataset to bring up the properties. You can also right-click and select Properties. Change the Query to the following code. It will now require that a year be provided when running the report:

```
SELECT SUM(TotalDue) AS TotalSales, MONTH(OrderDate) AS OrderMonth, YEAR(OrderDate) AS
OrderYear
FROM Sales.SalesOrderHeader
WHERE YEAR(OrderDate) = @Year
GROUP BY MONTH(OrderDate), YEAR(OrderDate);
```

When you run the report you are now required to type in a year before you see the results. Preview the report. Type in 2011 and click View Report. The totals displayed will be for only 2011 as shown in Figure 3-24.

Figure 3-24. *The results filtered by year*

The report is successfully filtered by the year. The user must already know which years are valid for the report. This is not too much of a problem for valid years, but what about departments or customers? In order to supply a valid list of years from which the user can choose, follow these steps:

1. Switch back to Design view and expand the Parameters folder in the Report Data window.

2. When you changed the query to filter by @Year, the Year parameter was automatically added. It is shown in Figure 3-25.

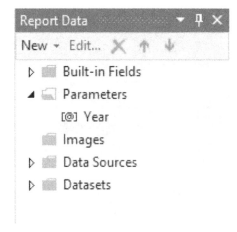

Figure 3-25. *The Year parameter*

3. Right-click Year and select Parameter Properties which brings up the Report Parameter Properties dialog box.

4. On the General page, change the Data type to Integer.

5. Switch to the Available Values page.

6. Change the Select from one of the following options to Get values from a query.

7. Change the Dataset property to Year.

8. Change the Value field to OrderYear.

9. Change the Label field to OrderYear. The dialog should look like Figure 3-26.

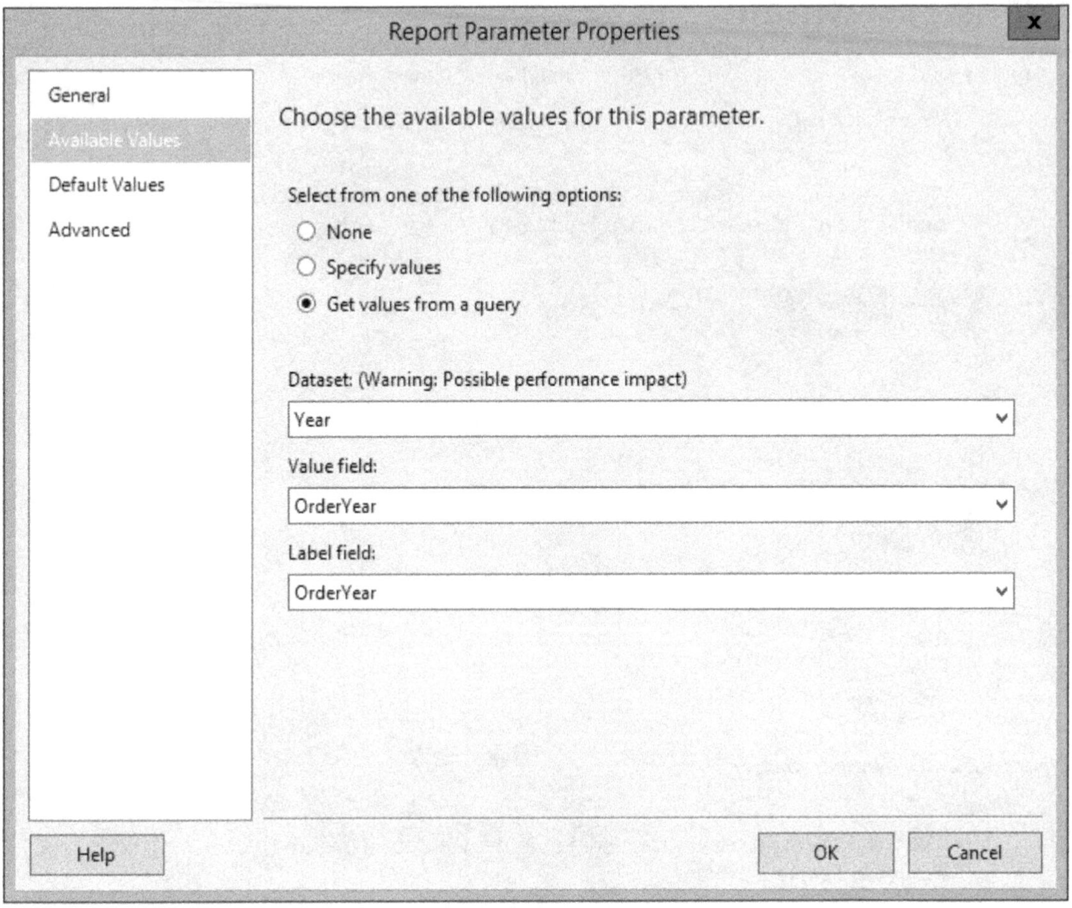

Figure 3-26. *The parameter properties*

    10.   Click OK to accept the properties.

You will learn much more about parameters in Chapter 6. For now, preview the report. You will see a dropdown list of valid years as shown in Figure 3-27.

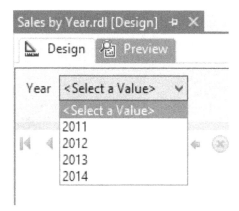

Figure 3-27. *The parameter list*

Experiment by running the report with different values for the Year parameter. The report data should change for each year chosen.

Summary

Understanding data sources and datasets is crucial to report development. Data sources are the connection strings to the database or other source of data. Datasets are the queries. It's possible to share both data sources and datasets among reports within a project. As a best practice, always share data sources. Share datasets if the query can be used across reports, such as for a commonly used parameter.

When you add a new report to a project remember that the following step is to add a data source to the report. The next step is to add a dataset.

Chapter 4 covers working with tables, text boxes, and other controls commonly used in reports.

CHAPTER 4

█ █ █

Working with Tables, Controls, and Report Sections

The past 15 years or so have been the age of the reality TV show. But long before "Extreme Makeover: Home Edition" and "American Idol," reality shows existed on public television. I remember watching unscripted programs about cooking, remodeling houses, and yoga as a child and young teen. My favorites were the shows about art. Theoretically, you could learn how to paint a masterpiece by watching the artist turn a blank canvas into a beautiful still life or landscape.

While reports may not be works of art, there are some similarities. You will start with a blank canvas, but while the data on the report is crucial, the look of the report is important as well.

In this chapter, you will learn more about tables and several other common controls used to build reports. You'll also learn how to add headers and footers to add features to make the report more professional and ready to publish.

Working with Tables

A report is made of several objects, some of which are linked to the dataset. Other objects give information to the viewer about the report, such as a title or page count. And some objects are there to give the report a certain look and feel such as a company logo. The most commonly used data region is the table control. Chapters 2 and 3 introduced tables. You linked tables to data and modified a few properties. In this section, you will learn even more about tables.

█ **Note** In Chapter 3, you learned how to create a project, data sources, and datasets. You also learned how to add a report to the project. Starting with this chapter, instead of telling you how to create these objects step by step, I'll just tell you what you need to create. Remember to click Add ➤ New Item when adding a new report to avoid starting the wizard. If you have trouble, refer back to Chapter 3.

To get started, follow these steps:

1. Launch SQL Server Data Tools (SSDT).

2. Create a new Report Server Project with the name Building Reports and solution name Beginning SSRS Chapter 4.

3. Add a new shared data source pointing to the AdventureWorks2016 database. Name it AdventureWorks2016.

© Kathi Kellenberger 2016
K. Kellenberger, *Beginning SQL Server Reporting Services*, DOI 10.1007/978-1-4842-1990-4_4

4. Create a new report with the name Working with Tables. Make sure that you add a blank report; do not start up the wizard. See Chapter 3 if you need help. The Solution Explorer should look like Figure 4-1.

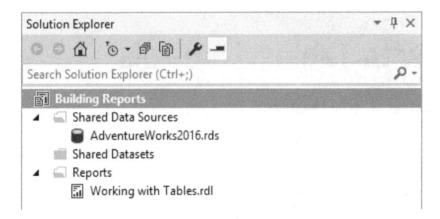

Figure 4-1. *The Solution Explorer*

■ **Note** The Always Show Solution property determines if the solution name is visible when there is only one project in the solution. See the section "Taking a Tour of SSDT" in Chapter 1 to learn how to set this option.

---

5. In the Report Data window, add a data source, AdventureWorks, that points to the shared data source AdventureWorks2016.

6. Create a new embedded dataset for the report named SalesSummary pointing to the AdventureWorks data source.

7. Type in the following query for the report:

```
SELECT YEAR(OrderDate) AS OrderYear, SUM(TotalDue) AS TotalSales
FROM Sales.SalesOrderHeader
GROUP BY YEAR(OrderDate);
```

8. The dataset properties should look like Figure 4-2. Click OK to create the dataset.

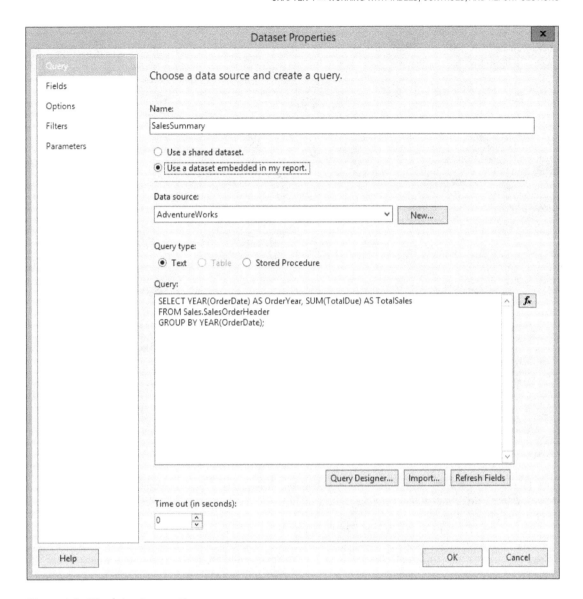

Figure 4-2. *The dataset properties*

9. Create a second embedded dataset called SalesDetails pointing to the AdventureWorks data source.

10. Add this query to the dataset and click OK.

```
SELECT CustomerID, SalesOrderID, OrderDate, TotalDue
FROM Sales.SalesOrderHeader;
```

The Report Data window should look like Figure 4-3.

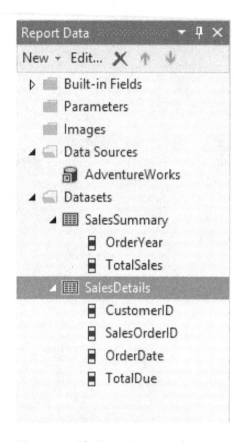

Figure 4-3. *The Report Data window*

Now that the datasets are in place, you can begin building your report. Add a new table to the design canvas. You can do this by dragging in a table control from the Toolbox or by right-clicking the canvas and selecting Insert ➤ Table.

Each table can be linked to just one dataset. In this case, you have two datasets. Once the first field is added, the table will be linked to the field's dataset. Hovering over a cell displays a small field list icon. If there is just one dataset, you will click the icon to bring up the field list. In this case, since there are two datasets, hover over the data source name to see both datasets as shown in Figure 4-4. Clicking the dataset name displays the list of fields.

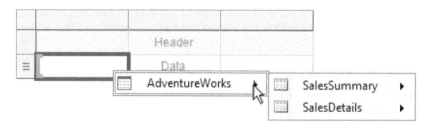

Figure 4-4. *Both datasets are available to the table*

Select OrderYear under SalesSummary. Now when you move to the next data cell, the SalesSummary fields are the only ones available. Choose TotalSales for the middle cell. You have an extra column on the right. Right-click the handle above the rightmost column and select Delete Columns to remove it.

Add a second table to the report under the first table. As you drag the second table around on the canvas, blue snap lines appear to help you make sure the objects are aligned. For this table, add the CustomerID, SalesOrderID, and OrderDate fields from the SalesDetails dataset to the report. By default, the table has three columns, but you have four fields to display. You can add additional columns by dragging the column to the right of the third column. When you see a blue bracket show up, drop the field. Figure 4-5 shows how this looks.

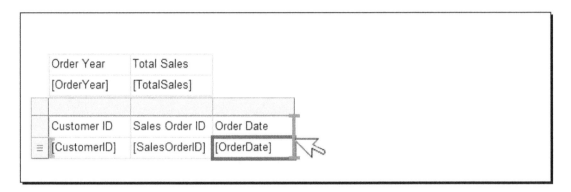

Figure 4-5. *Dragging a new field to the table*

You can also right-click on the third column and insert a column to the right.

Whenever fields are added to the data cells, the name of the cell is changed to the name of the field. A value fills in for the header cell as well adding spaces before the upper case letters. You can always edit the header cell if needed.

Property Window Properties

There are dozens of properties available for a table. You can set some of them in the Properties window. Select the intersection of the row and column handles of the first table to select the table. With the table selected, press F4 to open the Properties window. You can also click View ➤ Properties Window from the menu. At the top of the window shown in Figure 4-6, you will see that the table is not called a table at all. It is a *Tablix*.

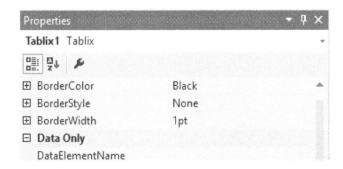

Figure 4-6. *The table is a Tablix*

■ **Note** A data region, such as a table, matrix, or list, is actually called a Tablix. Each one has a particular layout for displaying data. You can start with a table and turn it into a matrix by grouping on a column.

Properties in the Properties Window can be displayed either by categories of properties or alphabetically. Click the icons shown in Figure 4-7 at the top of the Property window to toggle back and forth. There is also an icon that resembles a wrench to open the Property Pages window. You'll learn more about the Property Pages in the next section.

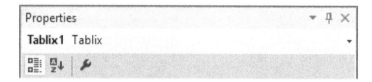

Figure 4-7. The property window icons

Since Tablix1 will be referred to several times in this chapter, give it a meaningful name: tblSalesSummary. To do this, change the Name property in the General category of the Properties window.

Look for the DataSetName property in the General category of properties. Notice that the property is set to SalesSummary.

As you click on each property, you will see a description of what the property does at the bottom of the Properties window. Most of the properties in the Properties window will not need to be changed, but whatever you change will affect the entire table. Change the BorderStyle Default property to Dotted and the BorderWidth Default property to 3 pt. The Properties window should look like Figure 4-8.

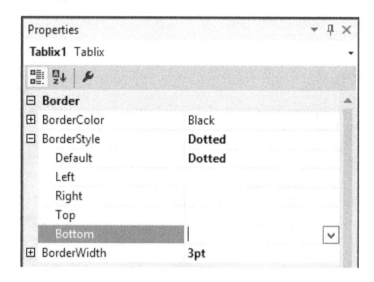

Figure 4-8. The modified properties

Preview the report. It should look like Figure 4-9.

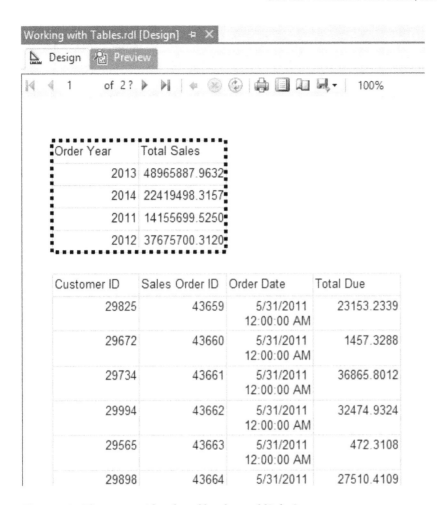

Figure 4-9. *The report with a dotted border on tblSalesSummary*

Property Dialog Properties

In addition to the Properties window, most report objects have a property dialog box as well. The properties available to change between the two methods overlap. Some of the more advanced properties, such as those involving interacting with other reports, are only available in the properties dialog. My opinion is that it is much easier to work in the dialog box when you can.

■ **Note** Every object added to the report has a name. By default the name will be the type of object plus a number. You may be wondering if you should give a descriptive name to every object. My rule is to name any object that will be referred to in an expression or calculation.

Switch back to design view and select the second Tablix. When the handles appear, right-click the intersection of the two handles. Select Tablix Properties as shown in Figure 4-10.

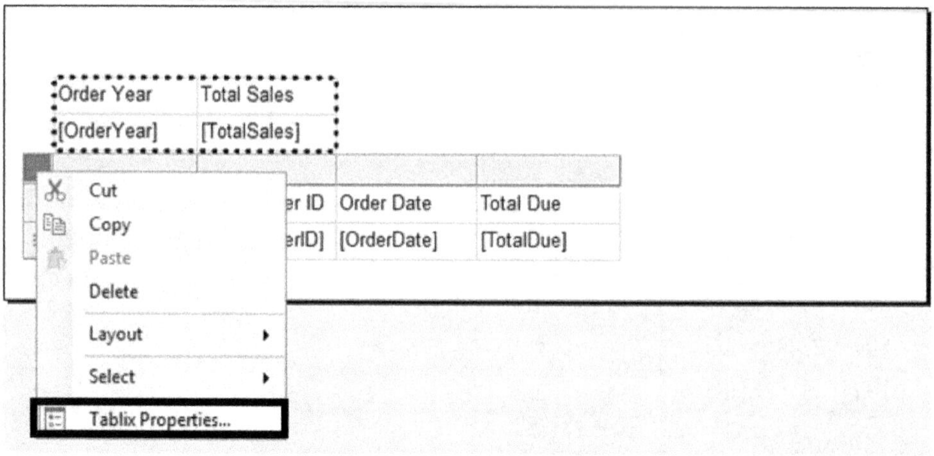

Figure 4-10. *Select Tablix Properties*

The Tablix Properties dialog has several pages. On the General page shown in Figure 4-11, change the Name to tblSalesDetails. Type Sales Details in the ToolTip property. Under Page break option, check Add a page break before.

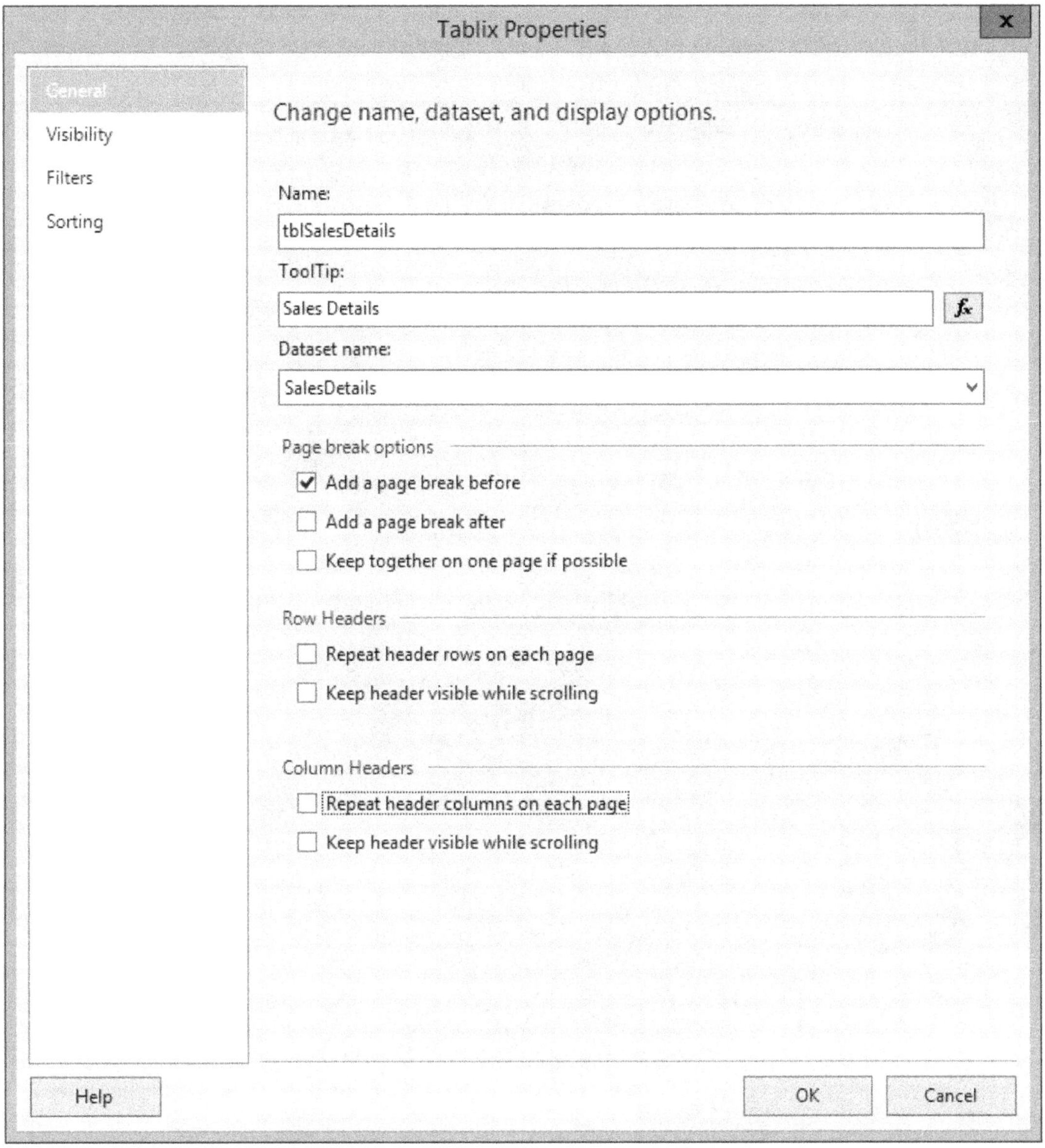

Figure 4-11. *The General page of the Tablix Properties*

Click OK to accept the property changes. Preview the report. Now tblSalesDetails will show up on the second page. Hover over the report to see the tooltip.

Take a look at Figure 4-11 once again. The properties shown in the Row Headers section to control the headers look like great ideas. They don't work, however. If you go back to design view, check these properties and view the report, you will see that they don't make any difference. The table headings will not be repeated on each page, and the header row will not be visible when scrolling. In Chapter 5, you will learn how to control the Row and Column headers.

Figure 4-12 shows the Visibility property page. You can choose to show or hide the table and to set the visibility of the table based on an expression. The value of a parameter might be a good expression to use.

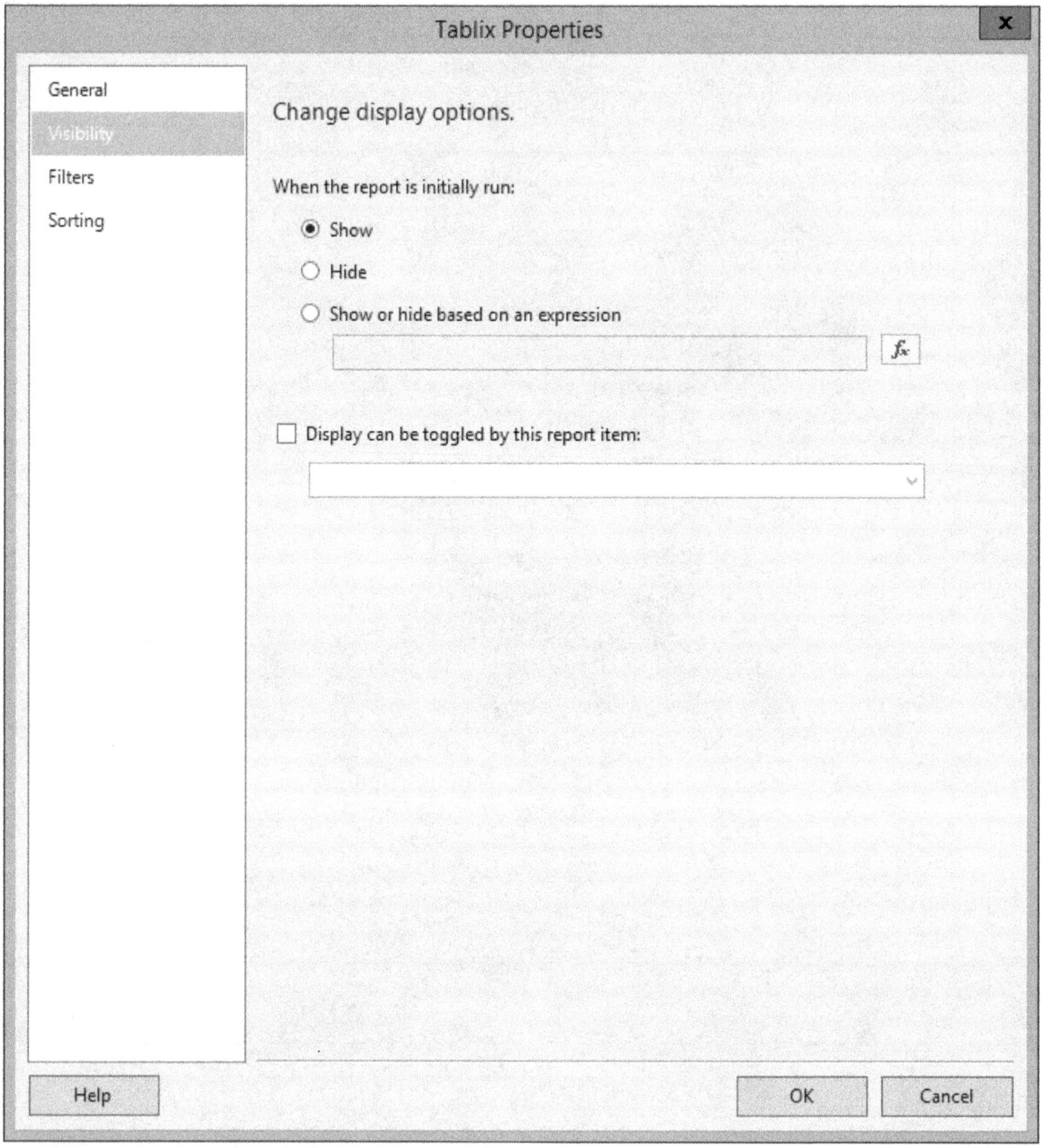

Figure 4-12. *The Visibility properties*

Theoretically, you can toggle the visibility by an item chosen from the dropdown list. If you check Display can be toggled by this report item and set the property to any of the items in the list, the report will display an error when run, so make sure to uncheck this before closing out of the dialog. The error occurs because all of the items in the list are part of the table and they can't be clicked before the report runs.

The Filters page allows you to add filters to the data returned by the dataset. Generally, this is a bad idea, because you should filter at the database. This capability is useful, however, if the source of data is not filterable, such as from an XML file. I did run across a use case a few years ago. I needed to use the same dataset, just with different filters, in several tables on the same report. Instead of pulling the data multiple times with different filters, I used one dataset and filtered at each table.

The most common reason I have found to bring up the Tablix Properties box is to change the display order of the data. The Sorting page is shown in Figure 4-13. Click Add and select CustomerID. Click Add again and select SalesOrderID. Save the properties and preview the report to see that the data is now sorted.

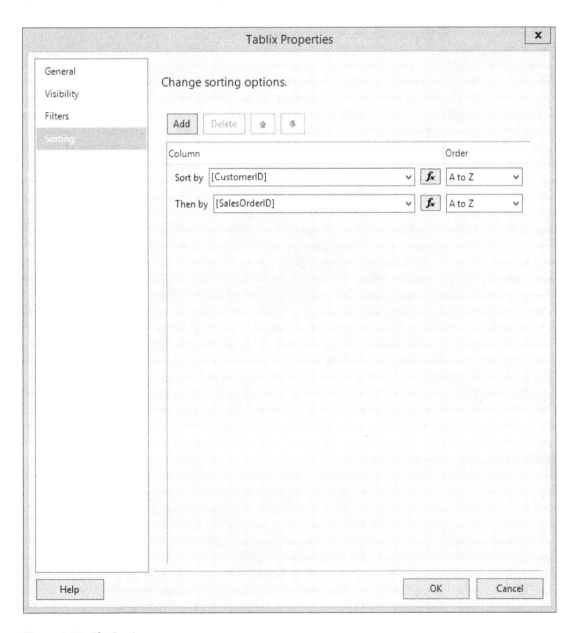

Figure 4-13. The Sorting page

Other Report Components

There are many other elements that you can add to reports to enable additional functionality or to make the report look more attractive and professional. Chapter 7 covers the visual elements such as charts and indicators, but you will take a look in this section at some of the more basic objects that you can add.

There are many reasons to add independent text boxes to reports (e.g., the report title and page number). Most of these objects should show up on each page. If you just add a text box to the top of the report body, it will show up only once. To make it show up on each page, include the information in a report page header.

Page Headers

Just like text boxes and tables, the page header has properties that you can modify to change the look or behavior. For example, you can add a background color or image. Follow these steps to add the page header to the report.

1. Right-click the report canvas and select Insert ➤ Page Header. You can also insert the page header from the Report menu.

2. Right-click the header and select Insert. The list of controls, as shown in Figure 4-14, you can add is very short. It's limited to controls that do not require being connected to data. Choose Text Box.

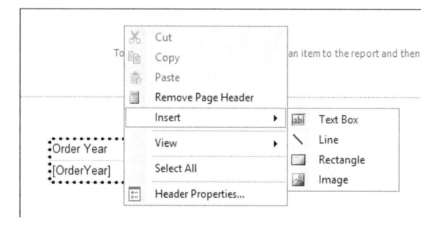

Figure 4-14. *The list of controls for the header*

3. From the Report Data window, expand Built-in Fields and drag Report Name to the text box. You can also drag fields directly to the header to add the text box automatically.

4. Click the edge of the text box so that the handle appears as shown in Figure 4-15.

Figure 4-15. *The handle on the text box*

5. By dragging the handle, reposition the text box so that it is located on the left side of the report.

6. Expand the text box so that it is as wide as the canvas.

7. With the text box selected, center the text. You can do this by clicking the Center icon in the menu or by changing the Horizontal property on the Alignment page of the Text Box Properties dialog.

8. Change the font size to 16 pt. Again, there are multiple ways to do this. Choose your favorite method: design menu, Properties window, or Property dialog. The text box will automatically grow vertically, but you can also increase the height.

9. Click the text box and add this text after the placeholder: `: Sales Details`

10. Select the words *Sales Details*. If you have the Properties window open, you will see that that the name of the currently selected object is Selected Text.

11. Italicize the words *Sales Details*. The text box should look like Figure 4-16.

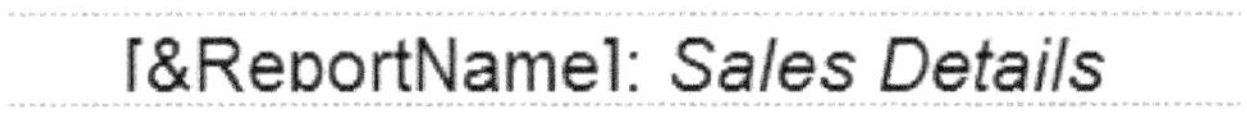

[&ReportName]: *Sales Details*

Figure 4-16. *Multiple formats in one text box*

12. Add a line control to the report header below the text box.

13. Adjust the line so that it is horizontal by dragging one end up or down.

14. Increase the size of the line by dragging each end.

15. Change LineWidth to 2pt in the Properties window. The LineWidth property is the thickness of the line. Lines are the simplest objects. They do not have a property dialog box.

16. Copy the line and paste in the header.

17. Move the second line so that it is close to and under the first line.

18. Right-click the report header and select Header Properties.

19. Uncheck Print on first page under Print options.

20. Click OK.

Now when you preview the report, you will see the header beginning with page 2 as shown in Figure 4-17. As you view subsequent pages, you will continue to see the page header.

Working with Tables: *Sales Details*

Figure 4-17. *Page 2 of the report*

Table Cell Formatting

Obviously, there is much you can do to make this report look better. To start, change the formats of the OrderDate and TotalDue fields by following these steps:

1. Return to design view.

2. Right-click the OrderDate cell and select Text Box Properties.

3. Select the Number page and choose Date under Category.

4. Select the 2000-01-31 format and click OK.

5. Bring up the Text Box Properties of the TotalDue cell.

6. Select the Number page and choose Currency under Category.

7. Make sure that Decimal places is set to 2.

8. Check use 1000 separator (,) and click OK.

When you preview the report, the table should look like Figure 4-18.

Customer ID	Sales Order ID	Order Date	Total Due
11000	43793	2011-06-21	$3,756.99
11000	51522	2013-06-20	$2,587.88
11000	57418	2013-10-03	$2,770.27
11001	43767	2011-06-17	$3,729.36
11001	51493	2013-06-18	$2,674.02
11001	72773	2014-05-12	$650.80

Figure 4-18. *The table with formatting*

The next step is to format the header row. Luckily, you do not need to format each cell individually. Follow these steps to format the row:

1. In design view, click a cell of tblSalesDetails.

2. When you do, the row and column handles appear. Select the handle next to the header row.

3. In the Properties window, change the BackgroundColor to CornflowerBlue.

4. Change the Font Color property to White.

5. Change the FontWeight property to Bold.

6. Now, click the Sales Order ID column.

7. Align each column to the right by selecting each one and then clicking the Align Right icon in the design menu.

8. The table may not line up well under the report header. Follow these steps to adjust the report.

9. Right click the report canvas and select View ➤ Ruler.

10. Drag the right edge of the report title so that it is 5 inches (12.5 cm) wide.

11. Select both lines. Hold down the CTRL key and press the right arrow until the lines are arranged under the title.

12. Drag the right edge of the report canvas to the left as far as it will go, which will be when it bumps up against a control.

To make long reports easier to read, you may want to alternate the background color of the rows. Unfortunately, there is no property that you can change to accomplish this. You can, however, use an expression to control the BackgroundColor property. Select the data row. In the Properties window, change the BackgroundColor to = IIf(RowNumber(Nothing) Mod 2 = 0, "LightBlue", "White").

Now preview the report. Page 2 of the report should look like Figure 4-19.

Working with Tables: *Sales Details*

Customer ID	Sales Order ID	Order Date	Total Due
11000	43793	2011-06-21	$3,756.99
11000	51522	2013-06-20	$2,587.88
11000	57418	2013-10-03	$2,770.27
11001	43767	2011-06-17	$3,729.36
11001	51493	2013-06-18	$2,674.02
11001	72773	2014-05-12	$650.80
11002	43736	2011-06-09	$3,756.99
11002	51238	2013-06-02	$2,535.96
11002	53237	2013-07-26	$2,673.06
11003	43701	2011-05-31	$3,756.99

Figure 4-19. The formatted Sales report

The ability to control the value of just about any property with an expression is a wonderful feature of SQL Server Reporting Services (SSRS). To know if a property can be controlled with an expression, look for the word Expression or the symbol *fx* in the property list or page.

Page Footers

A page footer is also quite useful for displaying information such as the page number and date the report was viewed. To add the page footer, follow these steps:

1. Switch to design view.

2. Right-click the report body and select Insert ➤ Page Footer.

3. Right-click the footer and select Footer Properties.

4. Uncheck the Print on first page property in the Print options section.

5. Click OK to accept the change.

6. From the Built-in Fields folder in the Report Data window, drag Execution Time into the footer.

7. Increase the width of the field so that it is about double the original width.

8. Add a text box to the footer. Be sure it lines up horizontally with the execution time text box.

9. Type this expression into the text box: Page [&PageNumber] of [&TotalPages]. You could also bring up the Expression dialog to help you write it.

▨ **Caution**   By adding the Total Pages property, the report must be completely built before the first page is returned to the user. If the report is extremely large, this could give the impression of a performance problem.

---

10. Double the width of the text box.

11. Add a line to the top of the footer.

12. Level the line and drag the edges to the width of the report.

When you run the report, the report footer on the second page should look like Figure 4-20.

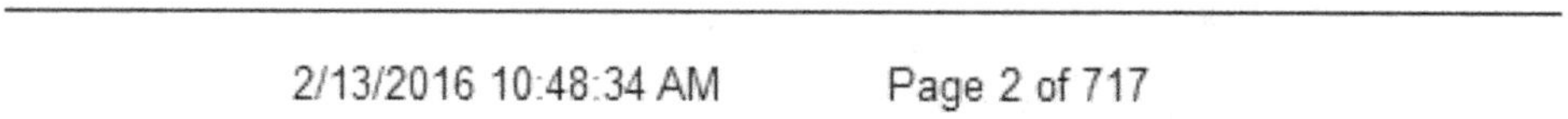

2/13/2016 10:48:34 AM   Page 2 of 717

Figure 4-20. *The report footer*

Report Cover Page

The first page of the report can serve as a cover page. You cannot add separate pages to the design canvas, but you can add page breaks before or after tables or some other objects. In this case, you have already turned on a page break before the tblSalesDetails table and turned off the page header and footer for the first page. In this example, you will add a rectangle control that will be a container for other controls. You will set the PageBreak property of the rectangle to force a page break.

To create the cover page, follow these steps:

1. In design view, drag the top of the footer down to increase the size of the report body to 7 inches (18 cm). Figure 4-21 shows the top of the footer, which is the dotted line.

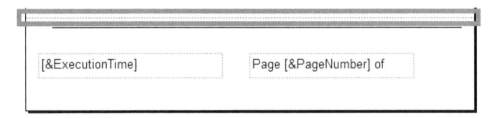

Figure 4-21. *The report footer*

2. Drag the tblSalesDetails table down so that it is just above the footer.

3. Delete the tblSalesSummary table.

4. Add a rectangle control to the report.

5. Expand the rectangle so that the top is right under the header, the bottom is right above the tblSalesDetail table, and the edges are at the width of the report canvas.

6. Right-click inside the rectangle and select Rectangle Properties. Change the Border properties to those shown in Figure 4-22.

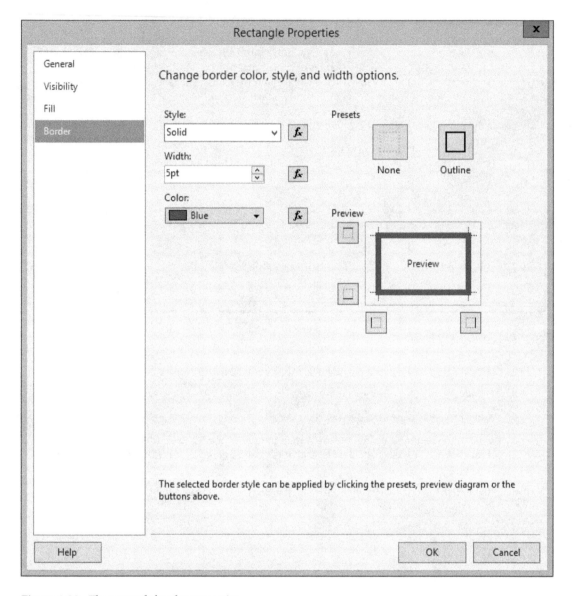

Figure 4-22. *The rectangle border properties*

7. On the General page, check Add a page break after. Make sure the Keep contents together on a single page, if possible is checked. Click OK to save the properties.

8. Bring up the Tablix Properties of the tblSalesDetails table, uncheck the Add a page break before property and click OK.

9. Add a textbox to the rectangle and type in Sales Report.

10. Change the font size to 20 pt.

11. Position the text box near the top and centered and resize it appropriately.

12. Add a table to the rectangle.

13. Fill in OrderYear and TotalSales from the SalesSummary dataset.

14. Remove the rightmost column.

15. Select the header row and bold the font.

16. Format the TotalSales cell as currency with no decimal places and with a 1000 separator.

17. Align each column on the right.

18. Change the table so that it sorts by OrderYear.

19. Add an image control to the top of the rectangle.

20. The Image Properties dialog will appear. Click Import and navigate to the AdventureWorks.jpg file located with the code for this chapter.

21. Click Ok to save the properties. Resize the image

This cover page is pretty simple. It has a blue outline, a title, and a summary of the data. For your report assignments, you could add a background color or graphic and more information. Figure 4-23 shows the top of the page cover.

Figure 4-23. *The cover page*

The page footer shows the first page of the details as page 2. Since you have a cover page, you may wish for that page to show as page 1. To fix this, follow these steps:

1. Switch back to design view.

2. Delete the pages text box.

3. Add a new text box to the page footer.

4. Right-click the new text box and select Expression.

5. Type in this expression:

   ```
   ="Page " & Globals!PageNumber -1 & " of " & Globals!TotalPages -1
   ```

6. Click OK.

Text Boxes with Data

You have seen that data regions are made of text boxes. You have also seen text boxes with disconnected information such as the report name in a text box. It is possible to populate an independent text box with data from one of the datasets. The text box cannot display detail data; it must display a summary value. To see this, drag the TotalSales field from the SalesSummary dataset to the report canvas. Take a look at the expression property as shown in Figure 4-24. You can find this by right-clicking the text box and choosing Expression.

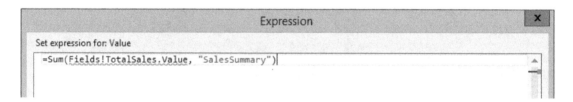

Figure 4-24. *The text box expression property*

The value has automatically been summarized with the Sum function. A second argument, SalesSummary, is the dataset context.

Calculated Fields

Reports often need calculated fields such as the sales margin. There are many ways to accomplish this. If the data originates in a database system, it is generally possible to add the calculation to the query itself. If that is not possible (e.g., the data source doesn't support it or the report developer is not able to modify the query), a calculation can be created within a cell. A problem arises, however, if the results of that cell need to be used in an expression in another cell. You can't nest a cell with an expression within another expression.

If a calculation is needed in multiple expressions, you can add a calculated field to the dataset. To learn how to add calculated fields, follow these steps:

1. From the Solution Explorer, add a new report named Calculated Field.

2. Add a data source to the report named AdventureWorks pointing to the AdventureWorks2016 shared data source. Review Chapter 3 if you need help with this step.

3. Add a new dataset named Sales pointing to the AdventureWorks data source with this embedded query:

```
SELECT TOP(1000) SOD.SalesOrderID, SOH.OrderDate,
    SOD.OrderQty, SOD.UnitPrice,
    P.StandardCost
FROM Sales.SalesOrderHeader AS SOH
JOIN Sales.SalesOrderDetail AS SOD
    ON SOH.SalesOrderID = SOD.SalesOrderID
JOIN Production.Product AS P ON P.ProductID = SOD.ProductID;
```

4. After saving the dataset, right-click and select Add Calculated Field which brings up the Fields page of the Dataset Properties dialog.

5. The dialog shown in Figure 4-25 displays the existing fields with a blank row at the bottom.

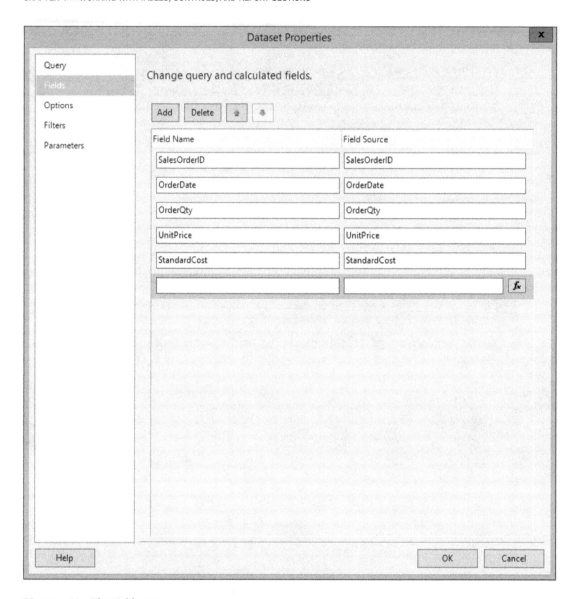

Figure 4-25. *The Fields page*

6.   Type in ExtendedCost as the first new Field Name.

7.   Click the *fx* icon to bring up the expression window.

8.   In the Category list, select Fields (Sales).

9.   In the Values list, double-click the OrderQty field.

10. Add a multiplication symbol (*) after the field name.

11. Double-click the StandardCost field. The final expression is

```
=Fields!OrderQty.Value * Fields!StandardCost.Value
```

12. Click OK to accept the expression.

13. Click Add ➤ Calculated Field to create another expression called ExtendedPrice. This expression is

```
=Fields!OrderQty.Value * Fields!UnitPrice.Value
```

14. Click OK to save the expression.

15. Click OK to save the new fields. You will see the new fields in the Sales dataset fields list as shown in Figure 4-26.

Figure 4-26. *The new calculated fields*

16. Add a table control to the report canvas.

17. Add the following fields to the table: SalesOrderID, OrderDate, OrderQty, StandardCost, ExtendedCost, UnitPrice, ExtendedPrice.

18. Add a new column to the table.

19. Right-click the new data cell and choose Expression.

20. Enter this expression:

```
=Fields!ExtendedPrice.Value - Fields!ExtendedCost.Value
```

21. Enter Margin in the header cell.

When you preview the report, you will see that the calculated fields work as expected.

The List Control

The list control works similarly to a table, but instead of cells, it has a freeform layout area that is repeated for each row of data. In this section, you will create a new report that uses the list control. Follow these steps to create the report.

1. Add a new report to the project named List Report.

2. Add a new data source to the report called AdventureWorks that is linked to the AdventureWorks2016 shared data source. Review Chapter 3 if you need help with this step.

3. Create a new dataset called ProductList pointing to the AdventureWorks data source with this embedded query:

```
SELECT P.ProductID, P.Name, PH.LargePhoto
FROM Production.Product AS P
JOIN Production.ProductProductPhoto AS PP ON P.ProductID = PP.ProductID
JOIN Production.ProductPhoto AS PH ON PP.ProductPhotoID = PH.ProductPhotoID
JOIN Production.ProductSubcategory AS SC ON SC.ProductSubcategoryID =
P.ProductSubcategoryID
JOIN Production.ProductCategory AS C ON C.ProductCategoryID =
SC.ProductCategoryID
WHERE PP.[Primary] = 1 AND C.ProductCategoryID IN (1,4);
```

4. Add a list control to the report canvas.

5. From the ProductList dataset, drag the ProductID and Name to the list. Text boxes are automatically created to hold the fields.

6. Double the width of the list.

7. Change the list's BorderStyle Default property to Solid.

8. From the Toolbox, drag an image control to the list.

9. This brings up the Image Properties dialog. Set the properties as shown in Figure 4-27.

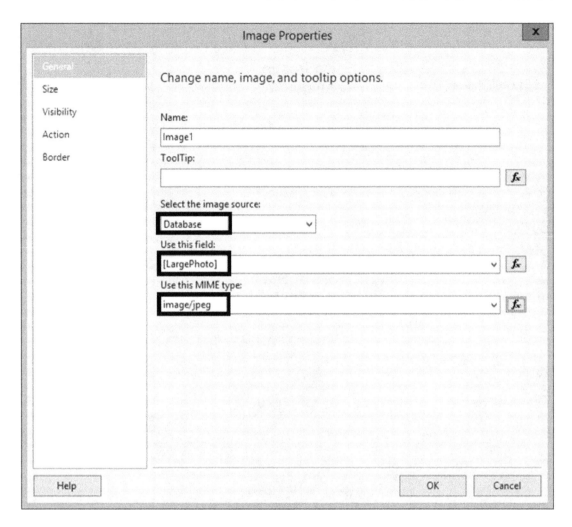

Figure 4-27. *The Image Properties box*

10. Click OK to accept the properties.

11. Expand the size of the image control.

Now when you preview the report as shown in Figure 4-28, you will see the images stored in the database for each product. Some of the products have images that just say "No Image Available." This is not a feature of SSRS when the image is missing; it's an actual image.

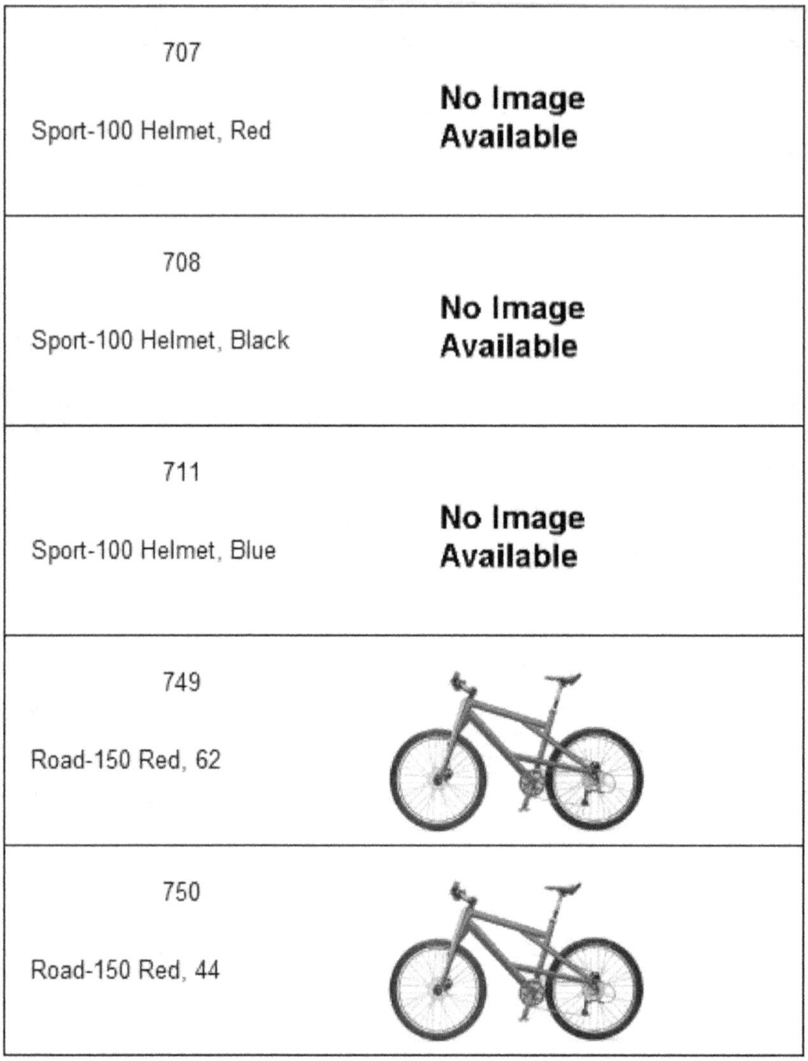

707 Sport-100 Helmet, Red	**No Image Available**
708 Sport-100 Helmet, Black	**No Image Available**
711 Sport-100 Helmet, Blue	**No Image Available**
749 Road-150 Red, 62	
750 Road-150 Red, 44	

Figure 4-28. *The list report*

Setting Report Properties

In Chapter 2, I told you about problems with printing reports. In some cases, the reports print with every other page blank. In this section, you will see how to solve this problem. Follow these steps to produce a report that is too wide to print:

1. In the design view of any report, enable the ruler by right-clicking the report canvas and clicking View ➤ Ruler.

2. Drag the right side of the report so that the report is about 9 inches (22.5 cm) wide.

3. Preview the report.

4. Switch to print layout view.

5. Scroll through the report one page at a time.

Because the report is wider than the default paper size, it prints a blank page after each populated page. If you have controls in place too far the right of the canvas, you will see the report split across pages as well. To solve this problem, you need to make sure that the report width does not exceed the paper where it will be printed.

Often, as the developer works on the report, adding and removing objects can cause the canvas to expand without notice. The first step is dragging in the right side of the report to remove any blank area.

Here are several things you can do to fix report printing.

-   Change the paper orientation from portrait to landscape.

-   Modify margins.

-   Change paper size. Make sure that this will match the paper you will use to print the report.

-   Adjust column widths.

Figure 4-29 shows the Report Properties dialog. It can be found in the Report menu.

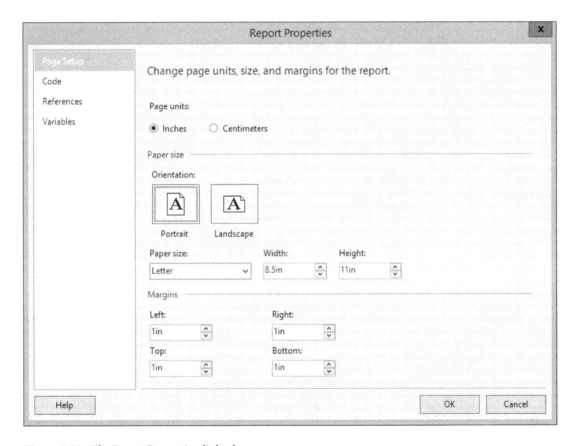

Figure 4-29. *The Report Properties dialog box*

As a rule of thumb, the width of the report plus the left and right margins should not exceed the width of the paper.

Summary

In this chapter you learned how to create a report from a blank canvas. There are many components that you can add to a report, including headers and footers. The values of almost all properties can be controlled with expressions, and you can add calculated fields to the dataset. It's always important to make sure that the report will fit on the printed page, and this chapter covered how to adjust the report properties to do so.

In Chapter 5, you will learn how to add grouping levels to reports.

■ ■ ■

Adding Grouping Levels to Reports

I had a birthday party when I turned six. I remember requesting chocolate cake with chocolate icing, chocolate ice cream, and chocolate milk. I also wanted my cake to be a multilayer wedding cake. I didn't get a wedding cake until years later at my actual wedding reception, and I'm sure the cake that my mom baked for my sixth birthday was a one-layer chocolate pan cake.

The reports you created in Chapters 3 and 4 displayed just one level in each table, much like that simple cake at my birthday party. In this chapter you will learn how to create reports with multiple grouping levels, including matrix reports.

Designing Your Report

The most important phase of any project, big or small, is the design. You must know what you are trying to accomplish before you start building the solution. A SQL Server Reporting Services (SSRS) report by itself is a relatively small project, but you will save time and frustration by understanding the requirements and coming up with a design before launching SQL Server Data Tools (SSDT).

Depending on the requirements for documentation at your company or department, you may end up just using a white board or paper to sketch out your design. For reports that require multiple grouping levels, calculations, or any advanced features, spend some time figuring out how the report should look. If you are lucky, the person requesting the report will provide a design layout along with the requirements.

Your company may have a very formal procedure to follow for new report requests with explicit requirements. Or, maybe a report request will made with just a phone call. In this chapter, you'll go through the process of building a report based on a simple request.

The Report Requirements

You have been given a report request from the sales department manager:

> We would like a report that shows sales by territory and year.

At this point, you could start building the report, but maybe you should ask some questions before you launch SSDT.

> How do I determine the territory? What fields need to show on the report? Do you need to see subtotals or other calculations? Should the report be grouped first by territory and then by year or the opposite? What is the lowest detail that needs to be shown on the report?

© Kathi Kellenberger 2016
K. Kellenberger, *Beginning SQL Server Reporting Services*, DOI 10.1007/978-1-4842-1990-4_5

When meeting with the manager, you find out the following:

- The query will be provided.

- Group first by year, then within the year, group by territory.

- For each year, show the total sales and the average sales over the territories.

- For each territory, display the territory name, ID, and total sales.

- Display store name, customer ID, and total sales as the lowest detail.

- Sort the detail by total sales in descending order.

- Sort the years by descending order.

- Sort the territories alphabetically.

You now have more specific requirements that may or may not be complete. In my experience, the person requesting the report will often have a better idea about what should be on the report and how it should be organized once he or she sees an actual report.

The Report Layout

Now that you have more details, you can begin to mock up the report layout on a white board or paper. You come up with the design shown in Figure 5-1.

Year	Average		Total
Territory ID	Territory Name		Total
	Customer ID	Store Name	Total

Figure 5-1. The report layout

The sales manager approves the layout, and you are ready to build the actual report.

Building a Report with Grouping Levels

This report will have three levels, but it will still be a relatively simple report. The skills you learn in this chapter will be easily applied to more complex reports.

■ **Note**   Chapter 3 covers working with data source and datasets. Chapter 4 covers adding controls and how to modify the properties. Refer to those chapters if you need step-by-step assistance with these topics.

Follow these steps to get started:

1. Using SSDT, create a new report project with the name Grouping Level Reports in a solution called Beginning SSRS Chapter 5.

2. Create a shared data source pointing to AdventureWorks2016 and also named AdventureWorks2016.

3. Add a new report to the project named Sales by Territory.

4. Add a data source named AdventureWorks to the report. The data source should point to the AdventureWorks2016 shared data source.

5. Add an embedded dataset to the report named SalesByTerritory that uses the AdventureWorks data source. Use this query:

```
SELECT YEAR(OrderDate) AS OrderYear, C.CustomerID, SUM(TotalDue) AS Sales,
    T.TerritoryID, T.Name AS Territory, s.Name AS Store
FROM sales.SalesOrderHeader AS SOH
JOIN Sales.SalesTerritory AS T ON SOH.TerritoryID = T.TerritoryID
JOIN Sales.Customer AS C ON SOH.CustomerID = C.CustomerID
JOIN Sales.Store AS S ON S.BusinessEntityID = C.StoreID
GROUP BY C.CustomerID, T.TerritoryID, T.Name,
    YEAR(OrderDate), S.Name;
```

6. Add a table to the report design canvas.

7. Add the CustomerID, Store, and Sales fields to the Data row.

The report design should look like Figure 5-2.

Customer ID	Store	Sales
[CustomerID]	[Store]	[Sales]

Figure 5-2. *The report design*

Add a Grouping Level to a Table Row

So far, the detail row is in place. The detail row will be nested inside a grouping level which is based on TerritoryID. There are two ways to add grouping levels, and you will learn both methods. Follow these steps to add a grouping level to a table:

1. Right-click any cell in the data row.

2. Select Add Group ➤ Row Group ➤ Parent Group as shown in Figure 5-3.

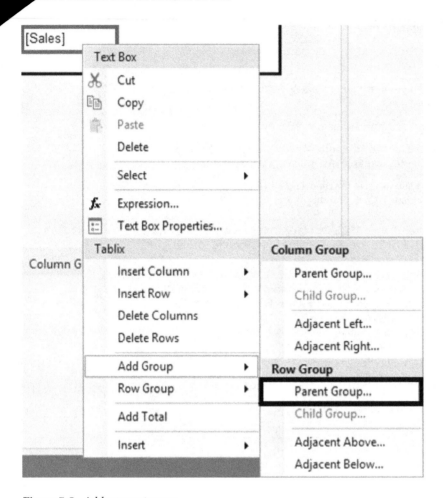

Figure 5-3. *Add a parent group*

3. This brings up the Tablix group dialog box. Select TerritoryID in the Group by dropdown box.

4. Select Add group header.

5. The dialog should look like Figure 5-4.

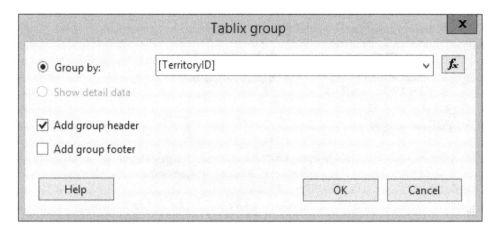

Figure 5-4. The Tablix group properties

6. Click OK. Figure 5-5 shows the table with the first grouping level.

Figure 5-5. The first group added to table

In this figure, the Sales data cell is selected. Notice that an orange bracket on the left of the CustomerID cell is visible. This indicates the grouping level of the selected cell. In this case, it is the detail level. Click one of the empty cells. Now the bracket can be found on the TerritoryID cell as shown in Figure 5-6.

Figure 5-6. The bracket is located on TerritoryID

This means that the cells in the middle row belong to the TerritoryID group. Add the Sales field to the empty cell on the right. Because this is a grouping level outside the details, the field is automatically summed as shown in Figure 5-7.

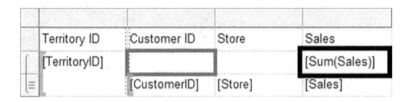

	Territory ID	Customer ID	Store	Sales
	[TerritoryID]			[Sum(Sales)]
		[CustomerID]	[Store]	[Sales]

Figure 5-7. *Sales is automatically summed*

In the CustomerID column, add the Territory field to the empty cell. The placeholder doesn't show it, but it is the first value found in the group. Since there is one Territory for each TerritoryID, this expression will work. Preview the report; it should look like Figure 5-8 at this point.

Territory ID	Customer ID	Store	Sales
	1 Northwest		14028557.7206
	29490	Bicycle Accessories and Kits	6972.9622
	29490	Bicycle Accessories and Kits	458.1082
	29497	Great Bikes	121846.1866
	29497	Great Bikes	299603.9787
	29497	Great Bikes	265207.1267

Figure 5-8. *The report preview*

Add a Grouping Level to the Grouping Window

The grouping window at the bottom of the report provides another way to add and configure grouping levels. Column Groups, on the right, are used for matrix reports. You'll learn about matrix reports in the section "Building a Matrix Report" later in this chapter.

At this point, you should see the TerritoryID row group in the window after switching back to design view. To add the next grouping level, follow these steps:

1. If the Grouping window is not visible below the report canvas, right-click the report and select View ➤ Grouping.

2. In the Row Groups window, click the down arrow next to TerritoryID.

3. Select Add Group ➤ Parent Group as shown in Figure 5-9.

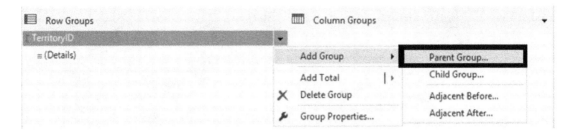

Figure 5-9. Add a Parent Group

4. Select OrderYear for the Group by property in the Tablix Group dialog.

5. Check Add group header.

6. Click OK. The Row Groups section will now look like Figure 5-10.

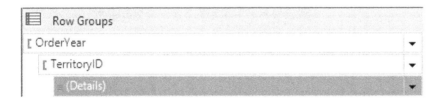

Figure 5-10. The row groups

7. In the empty cell in the Sales column, add Sales. This will be the total for the year.

8. In the empty cell in the Store column and second row, create an expression with the following formula: =Sum(Fields!Sales.Value)/CountDistinct(Fields! TerritoryID.Value). This will give the average sales over the territories, in other words, the average of the sum.

The report design should look like Figure 5-11.

Order Year	Territory ID	Customer ID	Store	Sales
[OrderYear]			«Expr»	[Sum(Sales)]
	[TerritoryID]	[Territory]		[Sum(Sales)]
		[CustomerID]	[Store]	[Sales]

Figure 5-11. The report design

When you preview the report, you will see that there is some cleanup to do. The report looks like Figure 5-12 at this point.

Order Year	Territory ID	Customer ID	Store	Sales
2011			1647825.270633 3333333333333 333	9886951.6238
	1		Northwest	2054523.7338
		29497	Great Bikes	121846.1866
		29521	Brightwork Company	62051.7757
		29531	Remarkable Bike Store	15396.7194
		29571	Moderately-Priced Bikes Store	16030.1459
		29580	Latest Sports Equipment	151771.5960

Figure 5-12. *The report in preview mode*

Formatting the Report

Follow these steps to begin formatting the report:

1. Switch to design view.

2. Format the four cells referring to the Sales field to currency with a thousands separator and no decimal points. Remember that once you format one cell, you can select the other three and paste in the format at one time.

3. Decrease the width of the Order Year and Territory ID columns.

4. Increase the width of the Store column.

The report is in the default group layout that you get when adding grouping levels to the report, but you can always rearrange it. You will now add two new rows to modify the layout. Follow these steps:

1. Right-click the TerritoryID cell and select Insert Row ➤ Inside Group ➤ Above. This adds a new row to the TerritoryID group.

2. Right-click the TerritoryID cell again and select Insert Row ➤ Outside Group ➤ Above. This adds a new row to the OrderYear group. The table layout should look like Figure 5-13.

Order Year	Territory ID	Customer ID	Store	Sales
[OrderYear]			«Expr»	[Sum(Sales)]
	[TerritoryID]			
		[Territory]		[Sum(Sales)]
		[CustomerID]	[Store]	[Sales]

Figure 5-13. *The report layout with two new rows*

3. In column 2, move the Territory ID heading to the cell above TerritoryID.

4. In column 3, move the Territory data field up one row.

5. Move the Customer ID heading above the CustomerID data cell.

6. Type the word Name above the Territory data cell. The layout should look like Figure 5-14 at this point.

Order Year			Store	Sales
[OrderYear]			«Expr»	[Sum(Sales)]
	Territory ID	Name		
	[TerritoryID]	[Territory]		
		Customer ID		[Sum(Sales)]
		[CustomerID]	[Store]	[Sales]

Figure 5-14. *The report preview after rearranging cells*

7. In column 4, move the Store heading to the cell above the Store data cell.

8. In the top row, above the average sales expression, type in Average Sales. The layout should look like Figure 5-15.

Order Year			Average Sales	Sales
[OrderYear]			«Expr»	[Sum(Sales)]
	Territory ID	Name		
	[TerritoryID]	[Territory]		
		Customer ID	Store	[Sum(Sales)]
		[CustomerID]	[Store]	[Sales]

Figure 5-15. *The layout after rearranging column 4*

119

9. In column 5, change the heading in row 1 to Total Sales.

10. In row 3, the territory header row, type in Total Sales.

11. In row 4, add the Sales field which will automatically sum.

12. Format the cell as currency with no decimals and with a thousands separator.

13. In row 5, the detail header row, change from the Sales field to the words Total Sales. The layout should look like Figure 5-16.

Order Year			Average Sales	Total Sales
[OrderYear]			«Expr»	[Sum(Sales)]
	Territory ID	Name		Total Sales
	[TerritoryID]	[Territory]		[Sum(Sales)]
		Customer ID	Store	Total Sales
		[CustomerID]	[Store]	[Sales]

Figure 5-16. *The layout after rearranging column 5*

14. Select the Average Sales column. Set the alignment to Align Right.

15. Select the Total Sales column. Set the alignment to Align Right.

16. Preview the report. It should look like Figure 5-17.

Order Year			Average Sales	Total Sales
2011			$1,647,825	$9,886,952
	Territory ID	Name		Total Sales
	1	Northwest		$2,054,524
		Customer ID	Store	Total Sales
		29497	Great Bikes	$121,846
		29521	Brightwork Company	$62,052

Figure 5-17. *The formatted report in preview mode*

You can use color and other formatting aspects to highlight the grouping levels. Follow these steps to format the groups:

1. Switch back to design view.

2. Select the top row and view the Properties window. Change the FontSize to 12 pt, the BackgroundColor to CornflowerBlue, and the Color property found in the Font section to White.

3. Repeat the formatting to the second row.

4. Select the third row. Change the BackgroundColor to LightBlue and FontWeight to Bold.

5. Repeat the formatting to the fourth row.

6. Select the fifth row. Bold the font.

Now when you preview the report, it looks like Figure 5-18.

Order Year				Average Sales	Total Sales
2011				$1,647,825	$9,886,952
	Territory ID	Name			Total Sales
	1	Northwest			$2,054,524
		Customer ID	Store		Total Sales
		29497	Great Bikes		$121,846
		29521	Brightwork Company		$62,052
		29531	Remarkable Bike Store		$15,397
		29571	Moderately-Priced Bikes Store		$16,030
		29580	Latest Sports Equipment		$151,772
		29582	Raw Materials Inc		$85,819
		29585	Parcel Express Delivery Service		$472
		29588	Nonskid Tire Company		$816

Figure 5-18. *The report preview with formatting*

As you scroll through the report, you will see that the headings are lost on subsequent pages. To enable the headings to repeat on each page, follow these steps:

1. Switch back to design view.

2. Click the down arrow located on the right of the Column Groups window and select Advanced Mode as shown in Figure 5-19.

Figure 5-19. *Select Advanced Mode*

3. When Advanced Mode is selected, the Row Groups and Column Groups will expand. You will see Static sections as shown in Figure 5-20.

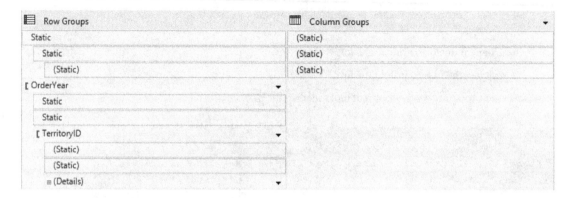

Figure 5-20. *The expanded grouping window*

4. Using the Properties window, change the RepeatOnNewPage value to True for each Static area found under Row Groups.

Now when you preview the reports, the headings will repeat on each page.

Sorting the Groups

There is one requirement that hasn't been addressed, sorting. The details must be sorted by the total sales in descending order. The report must also be sorted by year in descending order. Follow these steps to modify the sort order according to the requirements.

1. In the Row Groups window, right-click the Details group and select Group Properties.

2. In the Group Properties dialog, select the Sorting page.

3. Click Add.

4. Select Sales in the Sort by property.

5. Select Z to A in the Order property. The Group Properties dialog box should look like Figure 5-21.

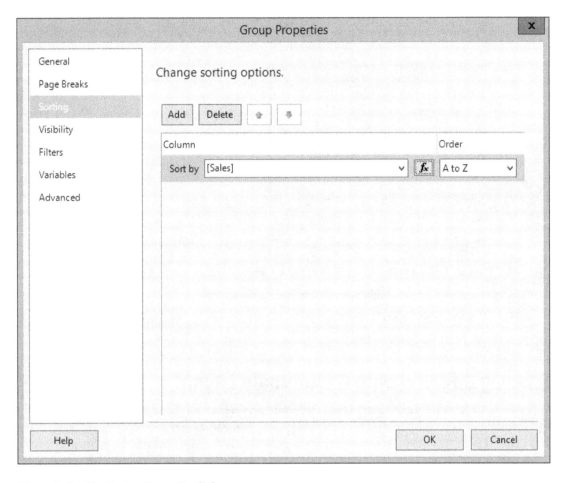

Figure 5-21. *The Sorting Properties dialog*

6. Click OK to accept the properties.

7. Modify the OrderYear group properties so that the report sorts in OrderYear descending order. The group automatically sorts by the grouping field.

8. Right-click the TerritoryID group and bring up the Group Properties.

9. Select the Sorting Page. Change the existing sort order from the TerritoryID field to the Territory field. Click OK when done.

Finalizing the Report

The report is almost ready to show the manager. Follow these steps to complete the report:

1. Switch to design view.

2. Drag the table to the top and left of the report.

3. Drag in the bottom and right edges of the report canvas.

4. From the Report menu, click Add Page Header and Add Page Footer.

5. Add a text box to the page header. Type in Territory Sales.

6. Expand the text box to the width of the canvas.

7. Center the text.

8. Increase the font size to 18 pt.

9. From Built-in Fields in the Report Data window, drag the Execution Time to the report footer.

10. Double the width of the text box.

11. Add another text box to the footer.

12. Add this expression to the text box:

    ```
    ="Page " & Globals!PageNumber & " of " & Globals!TotalPages
    ```

13. Launch the Report Properties dialog from the Report menu.

14. Change the Top and Bottom margins to 0.25 in or 0.635 cm.

15. View the report. Adjust any settings that make sense.

Now you are ready to show the report to the manager for feedback.

Building a Report with an Alternate Layout

After viewing the report, the manager verifies that the numbers are correct but would like a different layout. The alternate layout resembles the report you created with the wizard in Chapter 2 as shown in Figure 5-22.

First Report [Group]					
Region	**Order Year**	**Order Mont**	**Sales Order ID**	**Order Date**	**Total Due**
[Region]					[Sum(TotalDue
	[OrderYear]				Sum(TotalDue)]
		«Expr»			Sum(TotalDue)]
			[SalesOrderID]	[OrderDate]	[TotalDue]
			Total for [Group]		[Sum(TotalDue
					«Expr»

Figure 5-22. An alternate report layout

To create a report with this layout follow these steps:

1. Add a new report to the project named Sales by Territory 2.

2. Add a data source to the report named AdventureWorks pointing to the AdventureWorks2016 shared data source.

3. Add an embedded dataset to the report named SalesByTerritory pointing to the AdventureWorks data source with this query:

```
SELECT YEAR(OrderDate) AS OrderYear, C.CustomerID, SUM(TotalDue) AS Sales,
    T.TerritoryID, T.Name AS Territory, s.Name AS Store
FROM sales.SalesOrderHeader AS SOH
JOIN Sales.SalesTerritory AS T ON SOH.TerritoryID = T.TerritoryID
JOIN Sales.Customer AS C ON SOH.CustomerID = C.CustomerID
JOIN Sales.Store AS S ON S.BusinessEntityID = C.StoreID
GROUP BY C.CustomerID, T.TerritoryID, T.Name,
    YEAR(OrderDate), S.Name;
```

4. Add a table to the report canvas.

5. Add CustomerID, Store, and Sales to the Data row.

6. Add a parent row group to CustomerID.

7. On the Tablix group dialog, fill in TerritoryID as the Group by property.

8. Check Add group header. Click OK to accept the properties.

9. Add a parent row group to TerritoryID.

10. Fill in OrderYear as the Group by property.

11. Check Add group header. Click OK to accept the properties. The report should look like Figure 5-23.

Order Year	Territory ID	Customer ID	Store	Sales
[OrderYear]				
	[TerritoryID]			
		[CustomerID]	[Store]	[Sales]

Figure 5-23. *The report design with grouping levels*

12. Delete the Order Year and Territory ID columns. This does not delete the groups.

13. Right-click the Customer ID column and select Insert Column ➤ Left.

14. Repeat three more times so that there are four empty columns.

15. Type these values in the header row: Order Year, Average Sales, Territory ID, Name. The report design should look like Figure 5-24.

Order Year	Average Sales	Territory ID	Name	Customer ID	Store	Sales
				[CustomerID]	[Store]	[Sales]

Figure 5-24. *The report layout after adding headings*

16. The second row is part of the OrderYear group, but if you drag OrderYear to the cell underneath the Order Year heading, it will automatically sum since it is a numeric field. Type in [OrderYear] instead.

17. In the third row of the Territory ID column, type in [TerritoryID]. Again, make sure that this field doesn't sum the values.

18. Add the Territory field to the cell under Name in the third row.

19. Using the Expression dialog, add this expression to the cell under Average Sales in the second row:

```
=Sum(Fields!Sales.Value)/CountDistinct(Fields!TerritoryID.Value)
```

20. Add the Sales field to the two empty cells under the Sales header. These will automatically sum which is correct.

Now you have a grouping report with an alternate layout. When you preview the report, it should look like Figure 5-25.

Order Year	Average Sales	Territory ID	Name	Customer ID	Store	Sales
2011	1647825.27063 3333333333333 3333					9886951.6238
		1	Northwest			2054523.7338
				29497	Great Bikes	121846.1866
				29521	Brightwork Company	62051.7757
				29531	Remarkable Bike Store	15396.7194
				29571	Moderately-Priced Bikes Store	16030.1459
				29580	Latest Sports Equipment	151771.5960

Figure 5-25. *The report preview before formatting*

You are now ready to format the report. Follow these steps:

1. Format all cells involving Sales to currency with no decimal points and use a thousands separator.

2. Select the header row and view the Properties window. Change the BackgroundColor to CornflowerBlue.

3. Change the FontSize to 12 pt and the Color property in the Font section to White.

4. Decrease the width of Order Year, Average Sales, Territory ID, and Name columns.

5. Increase the width of the Store column.

6. Change the BackgroundColor of the second row to #8fb3f3.

7. Change the BackgroundColor of the third row to #c7d9f9.

8. Right-align the Sales column. The report design should look like Figure 5-26.

Order	Average	Territory ID	Name	Customer ID	Store	Sales
[OrderYe	«Expr»					[Sum(Sales)]
		[TerritoryID]	[Territory]			[Sum(Sales)]
				[CustomerID]	[Store]	[Sales]

Figure 5-26. *The report layout with formatting*

The report has very specific instructions for sorting each level. Follow these steps to sort the data:

1. Through the Group Properties dialog, change the sort order of the OrderYear group to descending.

2. Change the sort order of the detail row to Sales descending.

3. Change the sort order of the TerritoryID group to use Territory instead of TerritoryID.

The final set of tasks is to add the header and footer and modify the behavior of the headers. Follow these steps to complete the report:

1. Enable the Advanced Mode of the grouping window.

2. Set the RepeatOnNewPage to True in the Properties window for each Static level found in the Row Groups side.

3. Change the KeepWithGroup to After for each Static level found in the Row Groups.

4. To allow the header row to stay in place when scrolling down, select the top Static item.

5. Change the FixedData property to True.

6. Reposition the table so that it is in the top left corner.

7. Tighten the canvas by dragging in the edges.

8. Add a Report Page footer and Report Page header.

9. Add a text box to the header with the text Territory Sales.

10. Change the font size to 18 pt. Center the text box.

11. Add the Execution Time to the footer by dragging the field from the Built-in Fields folder in the Report Data window.

12. Add a text box with this expression to the footer:

```
="Page " & Globals!PageNumber & " of " & Globals!TotalPages
```

13. Change the report's top and bottom margins to 0.25in or 0.635cm.

Preview the report and try out the scrolling feature. Make any additional adjustments that are needed. The report should look like Figure 5-27.

Territory Sales

Order Year	Average Sales	Territory ID	Name	Customer ID	Store	Sales
2014	$1,316,752					$13,167,521
		9	Australia			$683,470
				29488	Nationwide Supply	$128,690
				30092	Cycle Parts and Accessories	$66,267
				29821	Bike Part Wholesalers	$59,272

Figure 5-27. *The Territory Sales report with an alternate layout*

If you scroll down, you should see that the header row stays in place.

Building a Report with a Space-Saving Layout

The alternate layout shown in the previous section may be preferred, but it takes more space than the default layout. There is a way to save some space, however. Follow these steps to create a report with a space-saving layout:

1. Right-click the Sales by Territory 2 report in Solution Explorer and select Copy.

2. Press CTRL + V to create a copy of the report.

3. Change the name to Sales by Territory 3.

4. Double-click the new report so that it is open in design view.

5. Right-click the second row and select Insert Row ➤ Inside Group – Below.

6. In the cell found in the third row under Order Year, type Territory ID.

7. Expand the Order Year column.

8. In the cell found in the third row under Average Sales, type Name.

9. In the cell found in the fourth row under Order Year, add [TerritoryID]. Type the placeholder to avoid summing the values.

10. In the cell found in the fourth row under Average Sales, add the Territory field.

11. Delete the Territory ID and Name columns. The report layout should look like Figure 5-28.

Order Year	Average	Customer ID	Store	Sales
[OrderYear]	«Expr»			[Sum(Sales)]
Territory ID	Name			
[TerritoryID]	[Territory]			[Sum(Sales)]
		[CustomerID]	[Store]	[Sales]

Figure 5-28. *The space-saving layout*

You can now decrease the width of the report canvas or add some additional fields to the report as needed. You may also want to modify the some of the formatting such as bolding the second row.

You can take advantage of the Padding property of the text box to indent the contents of a cell to create a hierarchical effect. Follow these steps to see how to do this:

1.   Align the first column to the left.

2.   Select the four cells related to the territory as show in Figure 5-29.

Order Year	Average	Customer ID	Store		Sales
[OrderYear]	«Expr»				[Sum(Sales)]
Territory ID	Name				
[TerritoryID]	[Territory]				[Sum(Sales)]
		[CustomerID]	[Store]		[Sales]

Figure 5-29. *Selecting the territory cells*

3.   Open the Properties window.

4.   Locate the Indent property in the Alignment category.

5.   Change the Left Indent property to 15 pt.

6.   Expand the width of the second column slightly. When you preview the report, it will look like Figure 5-30.

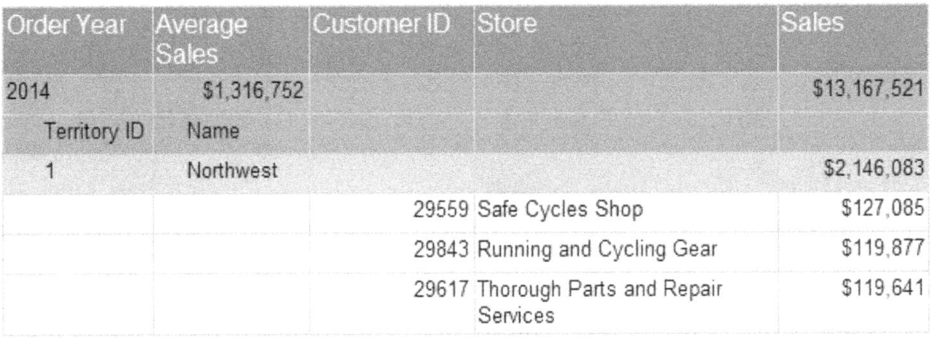

Order Year	Average Sales	Customer ID	Store	Sales
2014	$1,316,752			$13,167,521
Territory ID	Name			
1	Northwest			$2,146,083
		29559	Safe Cycles Shop	$127,085
		29843	Running and Cycling Gear	$119,877
		29617	Thorough Parts and Repair Services	$119,641

Figure 5-30. *The report with territory cells indented*

Building a Matrix Report

The manager is happy with the Sales by Territory 2 report but has another request. He would like you to create a matrix report that summarizes the sales by territory pivoted by the order year. Pivoting means that the data in the field will become a column header.

While a matrix report may seem daunting at first, it is really easy to create. Just like a regular report, spend some time figuring out the grouping levels. In this case, the row level group will be TerritoryID and the column level group will be OrderYear. You also need to determine the data field. This is the value that you wish to aggregate.

Figure 5-31 shows the matrix control before any cells are populated. Notice that the Data cell is the intersection of row and column groups.

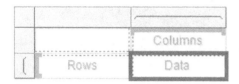

Figure 5-31. *The matrix control*

To create a matrix report, follow these steps:

1. Add a new report to the project named Sales by Territory Matrix.

2. Add a data source named AdventureWorks pointing to the AdventureWorks2016 shared data source.

3. Add an embedded dataset named SalesByTerritory pointing to the AdventureWorks data source with this query:

```
SELECT YEAR(OrderDate) AS OrderYear, C.CustomerID, SUM(TotalDue) AS Sales,
    T.TerritoryID, T.Name AS Territory, s.Name AS Store
FROM sales.SalesOrderHeader AS SOH
JOIN Sales.SalesTerritory AS T ON SOH.TerritoryID = T.TerritoryID
JOIN Sales.Customer AS C ON SOH.CustomerID = C.CustomerID
JOIN Sales.Store AS S ON S.BusinessEntityID = C.StoreID
GROUP BY C.CustomerID, T.TerritoryID, T.Name,
    YEAR(OrderDate), S.Name;
```

4. Drag a matrix control to the report.

5. In the Columns cell, add OrderYear. This is the pivoted data.

6. In the Rows cell, add TerritoryID.

7. In the Data cell, add Sales. It will automatically sum, which is what you need.

8. Right-click the first column and select Insert Columns ➤ Inside Group Right.

9. Add the Territory field to the new cell in the second row.

10. In the first row, change Territory heading to Name. The matrix layout should look like Figure 5-32.

Territory ID	Name	[OrderYear]
[TerritoryID]	[Territory]	[Sum(Sales)]

Figure 5-32. *The matrix report layout*

11. Format the data cell as currency with no decimal places and use a thousands separator.

12. Bold the first row.

13. Preview the report. It should look like Figure 5-33.

Territory ID	Name	2011	2012	2013	2014
1	Northwest	$2,054,524	$4,443,540	$5,384,411	$2,146,083
2	Northeast	$705,672	$3,272,240	$2,961,184	$873,896
3	Central	$1,126,646	$3,334,868	$3,371,278	$1,077,192
4	Southwest	$2,263,413	$7,976,842	$8,120,605	$2,471,177
5	Southeast	$1,847,745	$3,342,953	$2,700,616	$979,262
6	Canada	$1,888,952	$5,950,810	$6,310,918	$2,062,731
7	France		$1,166,211	$3,054,174	$977,725
8	Germany			$1,563,714	$717,890
9	Australia			$1,118,500	$683,470
10	United Kingdom		$1,126,623	$2,521,488	$1,178,095

Figure 5-33. The matrix report preview

14. Switch back to design view.

15. Right-click the Sum(Sales) cell and select Add Total ➤ Row.

16. Right-click the cell again and select Add Total ➤ Column. The report layout should look like Figure 5-34.

Territory ID	Name	[OrderYear]	Total
[TerritoryID]	[Territory]	[Sum(Sales)]	[Sum(Sales)]
Total		[Sum(Sales)]	

Figure 5-34. The matrix after adding totals

17. Add the Sales field to the empty cell under Total. It will automatically sum.

18. That cell will not pick up the formatting, so format it like the other Sales cells.

19. Bold the bottom row.

20. Bold the rightmost column. When you preview the report, it should look like Figure 5-35.

Territory ID	Name	2011	2012	2013	2014	Total
1	Northwest	$2,054,524	$4,443,540	$5,384,411	$2,146,083	**$14,028,558**
2	Northeast	$705,672	$3,272,240	$2,961,184	$873,896	**$7,812,991**
3	Central	$1,126,646	$3,334,868	$3,371,278	$1,077,192	**$8,909,983**
4	Southwest	$2,263,413	$7,976,842	$8,120,605	$2,471,177	**$20,832,038**
5	Southeast	$1,847,745	$3,342,953	$2,700,616	$979,262	**$8,870,575**
6	Canada	$1,888,952	$5,950,810	$6,310,918	$2,062,731	**$16,213,411**
7	France		$1,166,211	$3,054,174	$977,725	**$5,198,110**
8	Germany			$1,563,714	$717,890	**$2,281,604**
9	Australia			$1,118,500	$683,470	**$1,801,970**
10	United Kingdom		$1,126,623	$2,521,488	$1,178,095	**$4,826,207**
Total		**$9,886,952**	**$30,614,087**	**$37,106,887**	**$13,167,521**	**$90,775,447**

Figure 5-35. *The matrix after formatting the totals*

One of the interesting aspects of the table and matrix controls, is that you can start with one and, by modifying the groups, switch to the other. That is the reason that these controls are called Tablix. To change a matrix to a table, follow these steps:

1. Switch to design view.

2. Select the matrix and right-click the cross-section of the handle. Figure 5-36 shows where to right-click.

Territory ID	Name	[OrderYear]	Total
[TerritoryID]	[Territory]	[Sum(Sales)]	[Sum(Sales)]
Total		[Sum(Sales)]	[Sum(Sales)]

Figure 5-36. *Right-click the handle intersection*

3. Click Copy and then paste into the report canvas to create a copy of the matrix.

4. Select the new matrix and click right-click the OrderYear2 grouping level in the column grouping window.

5. Delete the group.

6. On the Delete Group dialog, select Delete Group and Related Rows and Columns. Click OK.

The matrix is now a table as shown in Figure 5-37. You can also start with a table and turn it into a matrix by adding a column group.

[TerritoryID]	[Territory]	[Sum(Sales)]
Total		[Sum(Sales)]

Figure 5-37. *The matrix is now a table*

Summary

Creating reports is an iterative process. You may start with a simple request and need to work with the requester to finalize the design. Or, you may start with an elaborate set of requirements including a detailed layout. Whenever you are creating a report with multiple grouping levels or complex features, be sure to sketch out the layout if a design is not given to you.

Most reports require one or more grouping levels. This chapter walked you through adding grouping levels and formatting the report using several techniques. You also created a simple matrix report. Be sure you understand how to add and configure groups before moving on to the next chapter.

In Chapter 6, you will learn how to make your report dynamic by adding parameters, linking reports, allowing the user to control sorting, and more.

CHAPTER 6

■ ■ ■

Making Reports Dynamic

If you have been following along with the demonstrations so far, you know how to create a nice formatted report with multiple grouping levels. Imagine that the department manager who requested reports in Chapter 5 would now like a separate report for each Order Year. Or maybe he or she would like to change the sort order or click a row to drill down to details.

In my career as a database administrator, I often created reports for some of the departments in the firm, and scenarios like this were all too common. I quickly learned to ask questions and anticipate what the requester might ask for next. I rarely created a report without providing options such as parameters.

In this chapter, you will learn how to make reports dynamic. It will save you time and make you look like a SQL Server Reporting Services (SSRS) rock star!

Adding Parameters to Report

Report developers most commonly use parameters to control the data that displays on the report. You saw in Chapter 3 that adding a parameter to the query in the dataset will automatically create a parameter in the report. You can also create parameters manually. In either case, there are a number of elements of the report that can change dynamically as the user runs the report.

In this section, you will start with a completed report and add a parameter to control the data displayed. To get started, follow these steps:

1. Launch SQL Server Data Tools (SSDT).

2. Create a new SSRS report project named Dynamic Reports in a solution named Beginning SSRS Chapter 6.

3. Create a shared data source named AdventureWorks2016 pointing to the AdventureWorks2016 database.

4. Right-click the Reports folder in the Solution Explorer and select Add ➤ Existing Item as shown in Figure 6-1.

© Kathi Kellenberger 2016
K. Kellenberger, *Beginning SQL Server Reporting Services*, DOI 10.1007/978-1-4842-1990-4_6

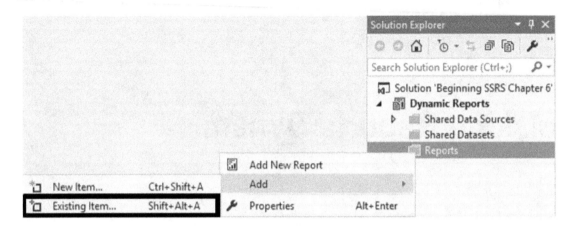

Figure 6-1. *Adding an existing report*

5. Navigate to the Sales by Territory Matrix.rdl file found in the project created in Chapter 5. If you did not create the report, you can use the report from the Code/ Download area of the Apress web site (Apress.com).

6. Click Add to import the report as shown in Figure 6-2.

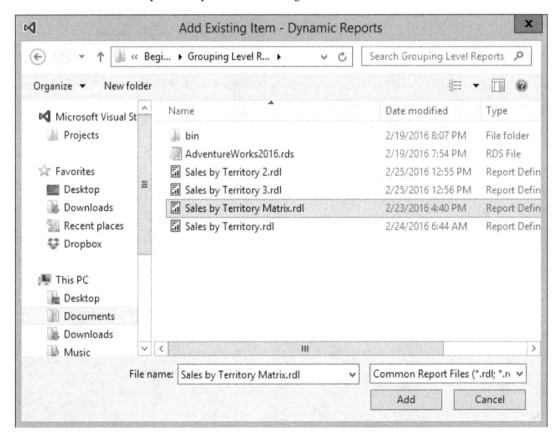

Figure 6-2. *Importing the existing report*

7. Double-click the report to open it in design view.

8. If there are two Tablix controls on the report, remove the table control and keep the matrix. Use Figure 6-3 as a guide.

Territory ID	Name	[OrderYear]	Total
[TerritoryID]	[Territory]	[Sum(Sales)]	[Sum(Sales)]
Total		[Sum(Sales)]	[Sum(Sales)]

[TerritoryID]	[Territory]	[Sum(Sales)]
Total		[Sum(Sales)]

Figure 6-3. *Remove the table control*

9. Preview the report to make sure it runs.

10. Switch back to design view.

11. In the Report Data window, open the properties of the SalesByTerritory dataset.

12. Change the query to

```
SELECT YEAR(OrderDate) AS OrderYear, C.CustomerID, SUM(TotalDue) AS Sales,
    T.TerritoryID, T.Name AS Territory, s.Name AS Store
FROM sales.SalesOrderHeader AS SOH
JOIN Sales.SalesTerritory AS T ON SOH.TerritoryID = T.TerritoryID
JOIN Sales.Customer AS C ON SOH.CustomerID = C.CustomerID
JOIN Sales.Store AS S ON S.BusinessEntityID = C.StoreID
WHERE YEAR(OrderDate) = @Year
GROUP BY C.CustomerID, T.TerritoryID, T.Name,
    YEAR(OrderDate), S.Name;
```

13. Click OK to save the change.

The difference between this query and the original is the WHERE clause. The query is now filtered by a @Year parameter. Adding a parameter to the query automatically adds a parameter to the report. Expand the Parameters folder in the Report Data window. The parameter should be visible as shown in Figure 6-4.

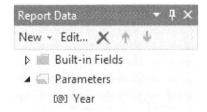

Figure 6-4. *The new parameter*

Now when you preview the report, you will be prompted to fill in a year. Try it out by entering several values. As long as you enter a whole number between 2011 and 2014, you will see data when you run the report.

To make sure that the person running the report supplies a valid value, you can provide a dropdown list from which the user can choose.

The Hard-Coded Parameter List

The parameter list can be a hard-coded list or from the results of a dataset. Follow these steps to create a list of years:

1. Right-click the Year parameter and select Parameter Properties.

2. Select the Available Values page.

3. Select Specify values.

4. Click Add.

5. Type in 2011 for both the Label and the Value.

6. Repeat for years 2012, 2013, and 2014. The dialog box should look like Figure 6-5.

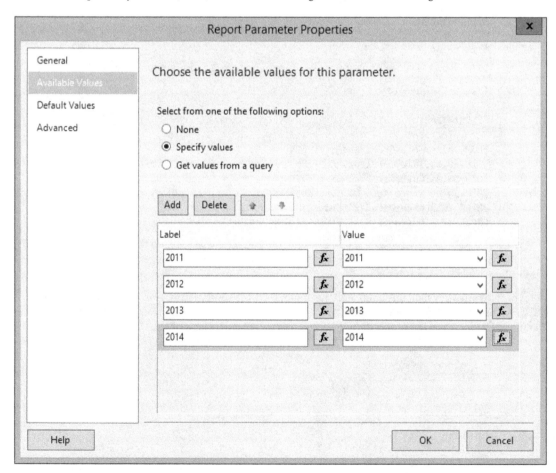

Figure 6-5. *Adding values to a parameter*

7. Click OK. Now when you preview the report, you will have a list to select from as shown in Figure 6-6.

Figure 6-6. *The parameter list*

The Label property of the parameter is what the end user sees; the Value property is what is passed to the query. In this case, they are the same.

Adding a Parameter List Based on a Query

Most of the time, it makes sense to base the parameter list on a query. This will save time since the list will not need to be manually maintained as the data changes. A parameter list will often be reused, so it makes sense to create a shared dataset instead of an embedded dataset. To create a parameter list based on a query, follow these steps:

1. Switch back to design view.

2. In the Solution Explorer, right-click on the Shared Datasets folder and select Add New Dataset.

3. Name the dataset Territory.

4. The Data Source property should point to AdventureWorks2016.

5. Make sure the Query type is set to Text.

6. Set the Query to

```
SELECT TerritoryID, Name AS Territory
FROM Sales.SalesTerritory;
```

7. The Shared Dataset Properties should look like Figure 6-7. Click OK.

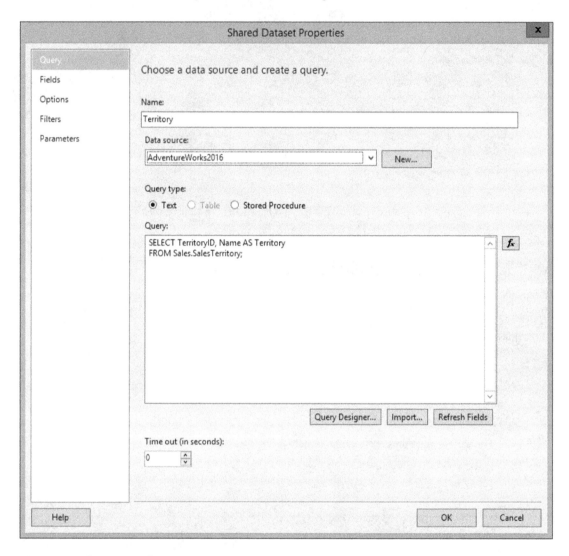

Figure 6-7. *The Territory dataset*

8. Add a new dataset to the report named Territory by using the Report Data window.

9. Choose Use a shared dataset.

10.   Select the Territory dataset from the dialog as shown in Figure 6-8.

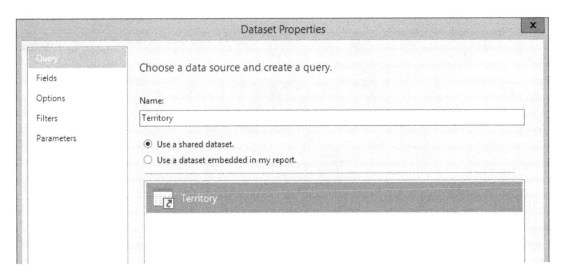

Figure 6-8. *The Territory dataset in the report*

11.   Click Ok.

12.   Bring up the properties of the SalesByTerritory dataset and change the query to

```
SELECT YEAR(OrderDate) AS OrderYear, C.CustomerID, SUM(TotalDue) AS Sales,
    T.TerritoryID, T.Name AS Territory, s.Name AS Store
FROM sales.SalesOrderHeader AS SOH
JOIN Sales.SalesTerritory AS T ON SOH.TerritoryID = T.TerritoryID
JOIN Sales.Customer AS C ON SOH.CustomerID = C.CustomerID
JOIN Sales.Store AS S ON S.BusinessEntityID = C.StoreID
WHERE YEAR(OrderDate) = @Year AND T.TerritoryID = @Territory
GROUP BY C.CustomerID, T.TerritoryID, T.Name,
    YEAR(OrderDate), S.Name;
```

13.   This will filter the dataset by both Year and Territory. Bring up the properties of the Territory parameter.

14.   Select the Available Values page.

15.   Choose Get values from a query.

16.   In the Dataset property, select Territory.

17.   In the Value Field property, select TerritoryID. This is what is needed in the query.

18. Select Territory for the Label Field property. This is what the end user will see. The dialog box will look like Figure 6-9.

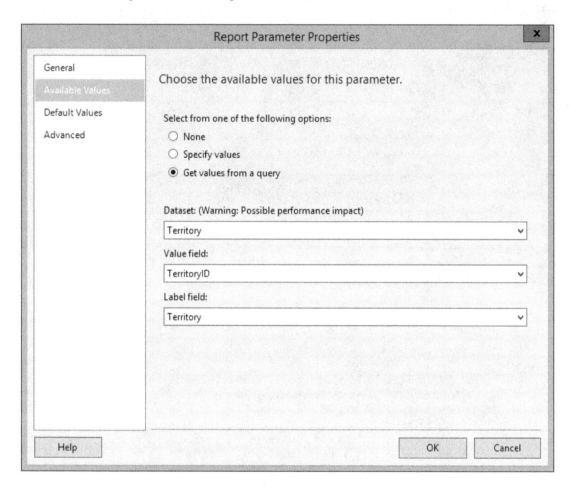

Figure 6-9. The Territory list properties

■ **Note** SSDT was released with a bug affecting shared datasets. Microsoft has promised to fix this in a later release. To correct the issue until then, close the project and navigate to the Territory.rsd file. Change `<Dataset>` to `<Dataset Name="Territory">`. Save the file and relaunch the project.

After you click OK to accept the parameter, preview the report. The new parameter should look like Figure 6-10.

Figure 6-10. *The Territory parameter list*

Run the report multiple times selecting different sets of parameters. You will see how the data displayed changes each time.

Default Parameters

Often, there is a set of parameters that are most likely to be chosen. To save the user time, you can add a default for each parameter. When launching the report, it will run once with the default parameters. The user may then select a different set of parameters and run it again if needed. To demonstrate this, follow these steps:

1. Switch to design view.

2. Bring up the properties of the Year parameter and select the Default Values page.

3. Select Specify values.

4. Click Add.

5. Fill in 2012. The dialog should look like Figure 6-11.

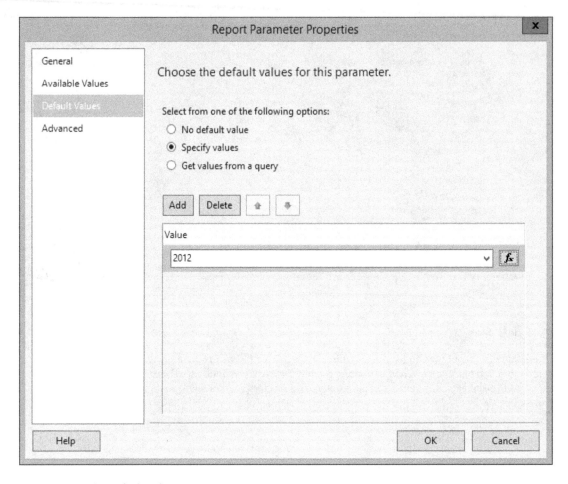

Figure 6-11. The Default Value properties

6. Click OK to accept the properties.

7. Add a default value to the Territory parameter using the same method.

8. Supply the value 7. The values must come from the TerritoryID field since that is what is passed to the query.

9. Click OK to accept the properties and preview the report. It should run without supplying parameter values and look like Figure 6-12.

Territory ID	Name	2012	Total
7	France	$1,166,211	**$1,166,211**
Total		**$1,166,211**	**$1,166,211**

Figure 6-12. The report after running with default parameters

Multivalued Parameters

So far, you have created parameters that allow the user to select one value from a list. It is also possible to create parameters that allow multiple selections. To do this, the query must be modified to select a list of values. Instead of `ColumnName = @Parameter`, you will use `ColumnName IN (@Parameter)`. Follow these steps to create a multivalued parameter:

1. Switch back to design view.

2. Bring up the properties of the Year parameter.

3. On the General tab, check Allow multiple values.

4. On the Default Values page switch back to No default values.

5. Click OK to save the changes.

6. Bring up the SalesByTerritory dataset.

7. To enable the expression in the WHERE clause to accept a list of values, the operator must change from equals (=) to IN. Change the Query property to

```
SELECT YEAR(OrderDate) AS OrderYear, C.CustomerID, SUM(TotalDue) AS Sales,
    T.TerritoryID, T.Name AS Territory, s.Name AS Store
FROM sales.SalesOrderHeader AS SOH
JOIN Sales.SalesTerritory AS T ON SOH.TerritoryID = T.TerritoryID
JOIN Sales.Customer AS C ON SOH.CustomerID = C.CustomerID
JOIN Sales.Store AS S ON S.BusinessEntityID = C.StoreID
WHERE YEAR(OrderDate) IN (@Year) AND T.TerritoryID = @Territory
GROUP BY C.CustomerID, T.TerritoryID, T.Name,
    YEAR(OrderDate), S.Name;
```

8. Run the report. The Year parameter should look like Figure 6-13.

Figure 6-13. *The multivalued parameter*

You can now select one or more years. Run the report a few times to see how the data changes when you choose different sets of parameters.

Cascading Parameters

Maybe you have noticed that some of the territories do not have sales in 2011. If you select 2011, all of the territories still show up in the parameter list. You can design a parameter so that it is based on the value of another parameter. Follow these steps to adjust the Territory parameter list based on the Year parameter.

1. Switch to design view.

2. Open the properties of the Territory dataset.

3. Switch to Use a dataset embedded in my report.

4. Set the Data source property to AdventureWorks.

5. Set the Query property to

   ```
   SELECT DISTINCT T.TerritoryID, T.Name AS Territory
   FROM sales.SalesOrderHeader AS SOH
   JOIN Sales.SalesTerritory AS T ON SOH.TerritoryID = T.TerritoryID
   JOIN Sales.Customer AS C ON SOH.CustomerID = C.CustomerID
   JOIN Sales.Store AS S ON S.BusinessEntityID = C.StoreID
   WHERE YEAR(OrderDate) IN (@Year);
   ```

6. Click OK to save the changes.

7. Open the Territory parameter properties.

8. On the Default Values page, change to No default value.

9. Click OK to save the changes.

Now when you preview the report, the Territory parameter will be grayed out until the year is chosen. Select 2011. The Territory parameter list will look like Figure 6-14.

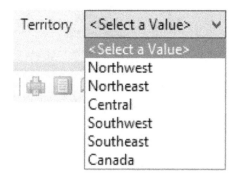

Figure 6-14. *The Territory parameter list based on 2011 sales*

You may have noticed that the query for the Territory parameter is similar to the SalesByTerritory query. Instead of querying for the values needed in the report, it produces a distinct list of territories. The tables used in the query are identical. The expression in the WHERE clause, YEAR(OrderDate) IN (@Year), is filtering on the Year parameter values chosen.

You can cascade even more parameters, each one based on the value of the previous parameter. Be cautious, however, that you do not negatively impact performance. For example, displaying the list of territories from the Sales.SalesTerritory table is much more efficient than the query used to display only those with sales in specifically chosen years. By the way, you can also select all years and the cascading will still work.

Parameter Placement

While in design view, you may have noticed a parameter section above the report. If you don't see it, right-click the design canvas and select View ➤ Parameters. The parameter section should look like Figure 6-15.

Figure 6-15. *The parameter section*

This feature is new with SQL Server 2016, and it provides improved flexibility with the parameter layout. In previous versions of SQL Server, you could change the order of parameters by selecting a parameter from the Parameters folder and clicking the up or down arrow as shown in Figure 6-16.

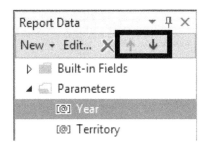

Figure 6-16. *The arrows for moving parameters*

Changing the positions using the method found in earlier versions of SSRS didn't always produce optimum placement. There was no control over which parameters end up on the same line. If you had a report with starting and ending dates, you may have wanted to keep the dates on the same line, but it wasn't always possible.

Beginning with SQL Server 2016, up to six parameters can display across. You can now control the placement of the parameters by dragging them to different cells. For example, drag the Territory parameter below the Year parameter. The parameter pane should look like Figure 6-17.

Figure 6-17. *The parameters rearranged*

When you preview the report, the parameters look like Figure 6-18.

Figure 6-18. *The parameters in preview mode*

By default, there are two rows for parameters. To increase this, right-click within the parameters area while in design view and select Insert Row Above or Insert Row Below. Be cautious, however, when moving parameters around if you are using cascading. Make sure that the parameters needed first are also displayed first to avoid a dependency error.

Parameter Data Types

So far, you have worked with numeric parameters, although the default type is text. You can also have parameters of other data types, and the parameter control will change based on the type. Follow these instructions to see the additional data types:

1. Make sure that the latest changes to the Sales by Territory Matrix report have been saved. You can click the Save icon or run the report.

2. Right-click Sales by Territory Matrix report and select Copy.

3. Right-click the project name, Dynamic Reports, and select Paste.

4. Change the name of the new report to Data Types.

5. While in design view, right-click the Parameters folder and select Add Parameter.

6. This brings up the Report Parameter Properties dialog box. Name the parameter DateParameter.

7. Fill in Date Parameter for the Prompt. The prompt is what the user will see when running the report.

8. Change the Data Type to Date/Time. The General page will look like Figure 6-19.

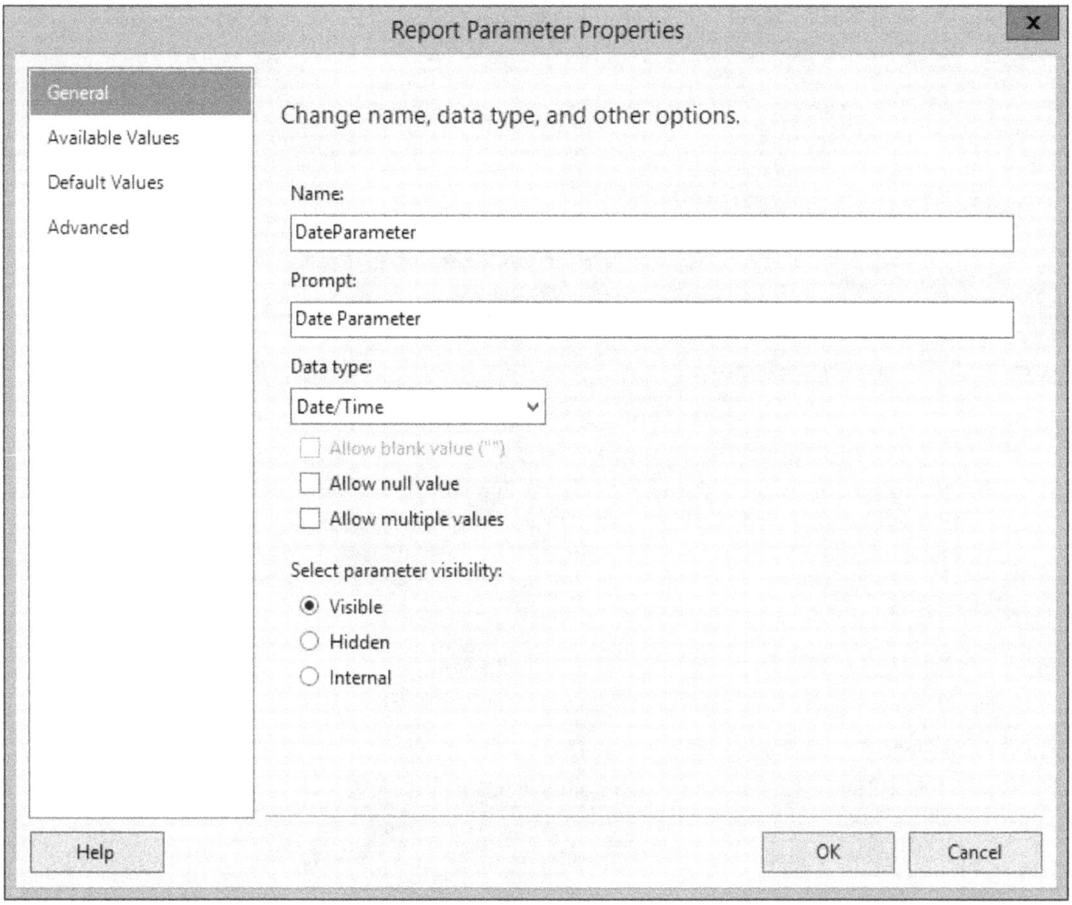

Figure 6-19. *The General properties of the parameter*

9. Click the Default Values page.

10. Select Specify Values.

11. Click Add.

12. Instead of typing in a value, click the *fx* symbol as shown in Figure 6-20 to open the expression dialog box.

Figure 6-20. *The expression button*

13. Fill in =Today() for the expression and click OK.

14. Click OK to accept the properties.

15. You should now see the Date Parameter in the grid with a calendar icon as shown in Figure 6-21.

Figure 6-21. *The new Date Parameter*

16. Preview the report.

17. If the Date Parameter is grayed out, select values for Year and Territory. The current date should be automatically chosen. You can either type in or use the calendar picker control to select another date. At least in the build I am using, the Date/Time parameter must be in the first position in order to be available before the other parameters are selected.

18. Switch back to design view and add another parameter.

19. Name this parameter TrueFalse with Prompt True or False.

20. Change the Data Type to Boolean and click OK.

21. This will add a set of radio buttons as shown in Figure 6-22.

Figure 6-22. *The new True or False parameter*

You can also set a default for the Boolean data type by adding True or False on the Default Values page.

The default data type is Text which lets the user type in anything. If you would like to restrict the input to a number, you can select the Integer or Float data types. The control will look like a regular text box but will accept only whole numbers or decimals, respectively.

Using Stored Procedures

SQL Server stored procedures, also called stored procs or sprocs, are objects that contain T-SQL scripts. These stored procs may contain programming logic such as loops, may update data, or may just contain a SELECT query. Stored procs are often used in SSRS datasets and they may require parameters by definition. If you are using a network SQL Server, you may need to contact your database administrator for rights to create a stored proc. Otherwise, to create a stored procedure, follow these steps.

1. Using SQL Server Management Studio (SSMS), connect to your SQL Server.

2. Click New Query which opens a query window.

3. Enter this code into the query window by typing or copying it from the Source Code/Download area for this book on the Apress web site (www.Apress.com). The code creates a stored proc that requires one argument, TerritoryID. It then returns the results filtered by that TerritoryID.

```
USE AdventureWorks2016;
GO
IF OBJECT_ID('usp_SalesByTerritory') IS NOT NULL
    DROP PROC usp_SalesByTerritory;
GO
CREATE PROC usp_SalesByTerritory @Year INT, @TerritoryID INT AS
    SELECT YEAR(OrderDate) AS OrderYear, C.CustomerID,
        SUM(TotalDue) AS Sales,
        T.TerritoryID, T.Name AS Territory, s.Name AS Store
    FROM sales.SalesOrderHeader AS SOH
    JOIN Sales.SalesTerritory AS T ON SOH.TerritoryID = T.TerritoryID
    JOIN Sales.Customer AS C ON SOH.CustomerID = C.CustomerID
    JOIN Sales.Store AS S ON S.BusinessEntityID = C.StoreID
    WHERE YEAR(OrderDate) = @Year AND T.TerritoryID = @TerritoryID
    GROUP BY C.CustomerID, T.TerritoryID, T.Name,
        YEAR(OrderDate), S.Name;
```

4. Click Execute in the menu bar or press F5 to create the proc as shown in Figure 6-23.

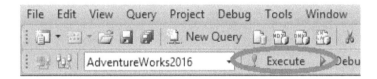

Figure 6-23. *The Execute icon*

5. Click New Query. In the new query window, test the proc by running the following code:

```
usp_SalesByTerritory @Year = 2011, @TerritoryID = 6;
```

Using a stored proc in a dataset is not very different from using a query. Follow these steps to create a report that uses the new stored proc.

1. Add a new report to the project named Stored Proc.

2. Add a data source pointing to AdventureWorks2016 and name it AdventureWorks.

3. Add an embedded dataset to the report named SalesByTerritory. Use the AdventureWorks data source.

4. Change the Query type to Stored Procedure.

5. When you do, the dialog changes. You will see a dropdown box listing the stored procedures in the database. Select usp_SalesByTerritory. The dialog should look like Figure 6-24.

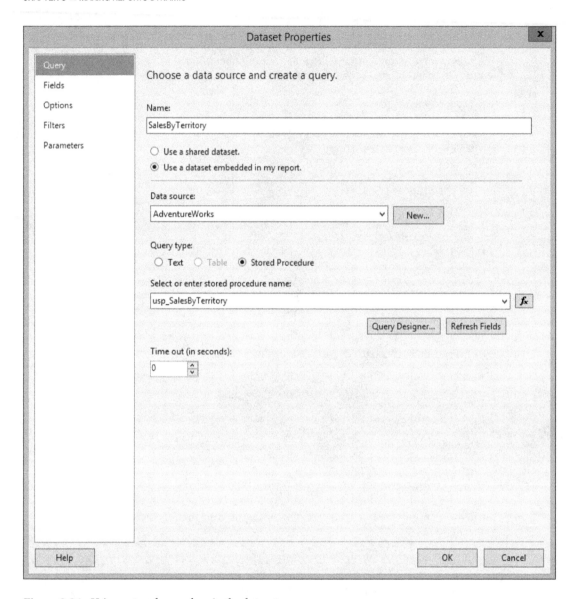

Figure 6-24. *Using a stored procedure in the dataset*

6. Click OK to create the dataset. It also adds the parameters to the report automatically.

7. Drag a matrix control to the report canvas.

8. Add OrderYear to the Columns cell.

9. Add Sales to the Data cell. It will automatically sum.

10. Add TerritoryID to the Rows cell. The layout should look like Figure 6-25.

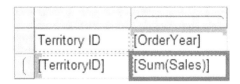

Territory ID	[OrderYear]
[TerritoryID]	[Sum(Sales)]

Figure 6-25. *The matrix design*

When you preview the report, you will be prompted for an OrderYear and TerritoryID. Fill in 2012 and 7 to see data.

There is an issue with using stored procedures, however. If you switch the parameters to accept multiple values, the report will no longer work. That's because the stored proc is expecting one integer value for each parameter, not a list of variables. To see the problem, follow these steps.

1. Switch back to design view.

2. Modify the Year parameter so that it allows multiple values.

3. On the Available Values page, select Specify values and add 2011 through 2014.

4. Click OK to save the changes.

5. Preview the report.

6. Select more than one year. Enter 7 for the TerritoryID.

7. When you click View Report, you will see an error message as shown in Figure 6-26.

An error occurred during local report processing.
An error has occurred during report processing.
Query execution failed for dataset 'SalesByTerritory'.
Error converting data type nvarchar to int.

Figure 6-26. *The error when multiple values are chosen for OrderYear*

As a workaround, the parameters for the stored proc must be changed to accept string values instead of just integers. Then, logic must be added to the proc to handle a comma delimited list instead of one value for each parameter. A function is often used to parse the list and save the values into a temporary table. Follow these steps to change the stored proc so that it can accept multiple values.

1. In SSMS, run this code to create a function to change a list of values into a table. You can copy the code from the Source Code/Download area for this book on the Apress web site (www.Apress.com).

```
USE AdventureWorks2016;
GO

IF OBJECT_ID('udf_ListToTable') IS NOT NULL
    DROP FUNCTION dbo.udf_ListToTable;
GO
```

153

```
CREATE FUNCTION dbo.udf_ListToTable(@List NVARCHAR(4000),
    @Delimiter NCHAR(1))
RETURNS @ValueList TABLE (ListItem NVARCHAR(50)) AS
BEGIN
    DECLARE @Pos INT;
    DECLARE @Item NVARCHAR(50);
    --Find the first delimiter
    SET @Pos = CHARINDEX(@Delimiter,@List);
    --loop until all items are processed
        WHILE @Pos > 0 BEGIN
            --insert the current item
            INSERT INTO @ValueList(ListItem)
            SELECT LEFT(@List,@Pos-1);
                --remove current item from the string
            SET @List = SUBSTRING(@List,@Pos+1,4000);
                --find the next delimiter
            SET @Pos = CHARINDEX(@Delimiter,@List);
END;
--add the last item
INSERT INTO @ValueList(ListItem)
SELECT @List;
RETURN;
END;

GO
```

2.   Click Execute to run the code and create the function.

3.   Click New Query to open another query window.

4.   Run this code in SSMS to modify the stored procedure.

```
USE AdventureWorks2016;
GO
IF OBJECT_ID('usp_SalesByTerritory') IS NOT NULL
    DROP PROC usp_SalesByTerritory;
GO
CREATE PROC usp_SalesByTerritory
    @YearList NVARCHAR(4000), @TerritoryIDList NVARCHAR(4000) AS

        DECLARE @Years TABLE (OrderYear INT);
        DECLARE @Territories TABLE(TerritoryID INT);
      --Save the lists into table variables
        INSERT INTO @Years(OrderYear)
        SELECT ListItem
        FROM dbo.udf_ListToTable(@YearList,',');

        INSERT INTO @Territories(TerritoryID)
        SELECT ListItem
        FROM dbo.udf_ListToTable(@TerritoryIDList,',');
```

```
--Change the query to use IN lists in the WHERE clause
SELECT YEAR(OrderDate) AS OrderYear, C.CustomerID,
        SUM(TotalDue) AS Sales,
    T.TerritoryID, T.Name AS Territory, s.Name AS Store
FROM sales.SalesOrderHeader AS SOH
JOIN Sales.SalesTerritory AS T ON SOH.TerritoryID = T.TerritoryID
JOIN Sales.Customer AS C ON SOH.CustomerID = C.CustomerID
JOIN Sales.Store AS S ON S.BusinessEntityID = C.StoreID
WHERE YEAR(OrderDate) IN (SELECT OrderYear FROM @Years)
        AND T.TerritoryID IN (SELECT TerritoryID FROM @Territories)
    GROUP BY C.CustomerID, T.TerritoryID, T.Name,
    YEAR(OrderDate), S.Name;

GO
```

5. Click Execute to run the code.

You now have a function that can be used whenever you run into this problem. The stored procedure uses the new function to insert each list into a table variable. Then, in the WHERE clause, the table variables are used as IN lists. If your source of data is hosted on SQL Server 2016, you can use a new built-in function called STRING_SPLIT() instead of the custom function used in this example.

Now it is time to make a few changes to the report to work with the changes to the stored proc. Follow these steps:

1. Switch to design view of the Stored Proc report.

2. Bring up the properties of the SalesByTerritory dataset.

3. On the Query page, click Refresh Fields. This will create new parameters that match the updated stored procedure, @YearList and @TerritoryIDList.

4. Click OK.

5. Expand the Parameters folder.

6. Delete the Year and TerritoryID parameters. They are no longer needed with the modified stored proc.

7. Create a new dataset named Territory. It should point to the shared Territory dataset.

8. Bring up the properties of the TerritoryIDList parameter.

9. Check Allow multiple values.

10. On the Available Values page, select Get values from a query.

11. Select Territory from the Dataset list.

12. Select TerritoryID from the Value field list.

13. Select Territory from the Label field list.

14. Click OK to accept the changes.

15. Open the properties of the YearList parameter.

16. Check Allow multiple values.

17. On the Available Values page, enter 2011 through 2014 as hard-coded values.

18. Click OK to save the properties.

19. Drag the parameters to the left in the parameter pane. The two deleted parameters left empty spaces.

20. Preview the report; it should now work as expected when multiple values are chosen.

Stored procedures are often the preferred type of command when working in SSRS. They allow reuse of code and also can be created to be more secure than just allowing the user to run queries against the tables. The main downfall is that the report developer may not have rights to create or modify stored procs in the database.

Controlling Properties

While filtering the data is a very common reason to make a report dynamic, you can also control almost any property dynamically. To do this, you will take advantage of expressions based on the data or parameters. You saw an example of this in Chapter 4 where you set up alternating background colors.

You can control any property that shows Expression as a property choice or with the *fx* icon next to the property. This section covers several examples.

Visibility

You can show or hide objects such as tables or rows with an expression, but the value of the expression must be known before the report actually displays. Follow these steps to use a parameter to control visibility.

1. Add a new report to the project named Visibility.

2. Add two rectangle controls to the report.

3. Change the Fill Color of the two rectangles to a dark color such as Blue.

4. Alter the sizes so that one is a large rectangle and one is a small square. The report layout should look like Figure 6-27.

Figure 6-27. *The report with two rectangles*

5. Add a parameter to the report named ShowRectangle with the Prompt property set to Show Rectangle.

6. Change the Data type to Boolean.

7. Click OK to create the parameter.

8. Add a second parameter named ShowSquare with the Prompt property set to Show Square.

9. Change the Data type to Boolean.

10. Click OK to create the second parameter.

11. Right-click the rectangle shape and select Rectangle Properties.

12. Switch to the Visibility page.

13. Change the When the report is initially run property to Show or hide based on an expression.

14. Click the *fx* icon to open the Expression dialog. Notice that you will be setting an expression to hide the rectangle as shown in Figure 6-28. This may seem counter-intuitive, so always keep it in mind when creating an expression for this property.

Figure 6-28. Set the expression for Hidden

15. In the Category list, select Parameters. You will see the two parameters listed in the Values list as shown in Figure 6-29.

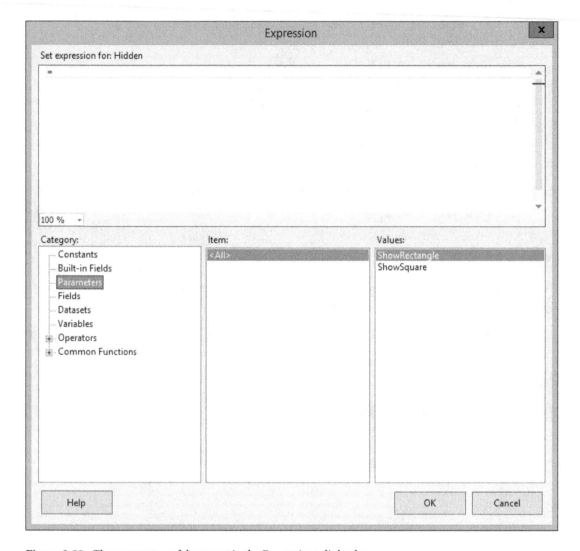

Figure 6-29. *The parameters of the report in the Expressions dialog box*

16. Double-click ShowRectangle.

17. If the expression evaluates to True, the rectangle will be hidden. Change the expression to

    ```
    =Not Parameters!ShowRectangle.Value
    ```

18. Click OK to dismiss the Expression dialog and then OK to dismiss the Rectangle Properties dialog.

19. Repeat the process for the small square shape, selecting the ShowSquare parameter.

20. Preview the report to test it. Figure 6-30 shows an example.

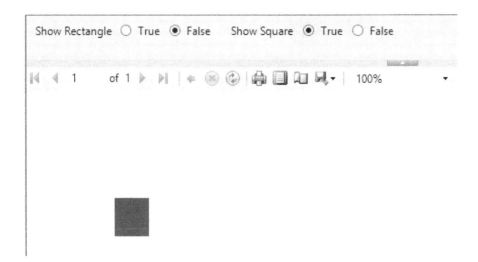

Figure 6-30. *The report showing just the square*

Another way to control visibility is often used with groups. This was demonstrated in the wizard report in Chapter 2 where you could click a plus sign to expand the details.

You will set a property on a group to toggle the visibility based on a cell found in the parent group. To learn how to do this manually, follow these steps:

1. Import the Sales by Territory 2.rdl file from the Chapter 5 project by right-clicking Reports and selecting Add ➤ Existing Item.

2. Double-click the report to open it in design view.

3. Right-click the TerritoryID cell and select Text Box Properties.

4. Change the Name property to TerritoryID if it is not set that way already.

5. Click OK to accept the change.

6. In the Row Groups window, right-click the Details group. This is the innermost child group.

7. Select Group Properties.

8. Select the Visibility page.

9. Select Hide and check Display can be toggled by this report item.

10. In the dropdown list, select TerritoryID. The properties should look like Figure 6-31.

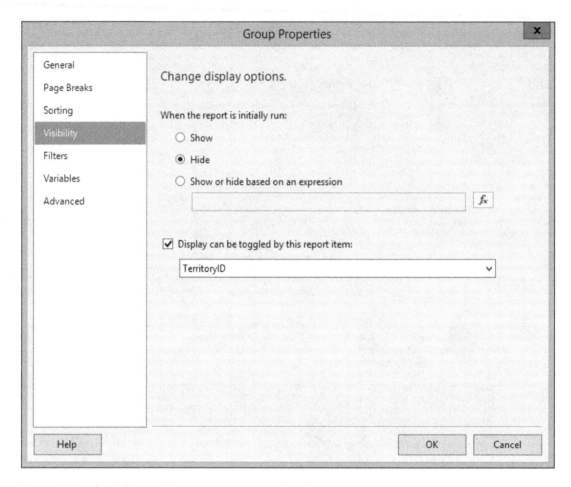

Figure 6-31. *The visibility settings*

11. Click OK to accept the properties.

12. Make sure that the OrderYear cell is named OrderYear as was done with TerritoryID.

13. Right-click the TerritoryID group and select Group Properties.

14. On the Visibility page, select Hide and make this group toggle by OrderYear.

15. Click OK to accept the properties. You do not need to set the property on the OrderYear group since it is the highest level parent.

16. Preview the report. It should look like Figure 6-32.

Territory Sales

Order Year	Average Sales	Territory ID	Name	Customer ID	Store	Sales
⊞ 2014	$1,316,752					$13,167,521
⊞ 2013	$3,710,689					$37,106,887
⊞ 2012	$3,826,761					$30,614,087
⊞ 2011	$1,647,825					$9,886,952

2/25/2016 12:29:43 PM   Page 1 of 1

Figure 6-32. *The report with collapsed sections*

When you click the plus sign, next to the year, you will see the territories for that year. When you click the plus sign next to the TerritoryID, you will see the details for that territory as shown in Figure 6-33.

Territory Sales

Order Year	Average Sales	Territory ID	Name	Customer ID	Store	Sales
⊟ 2014	$1,316,752					$13,167,521
		⊟	9 Australia			$683,470
				29488	Nationwide Supply	$128,690
				30092	Cycle Parts and Accessories	$66,267
				29821	Bike Part Wholesalers	$59,272
				30097	Helpful Sales and Repair Service	$56,467
				29823	Popular Bike Lines	$54,683
				29628	Budget Toy Store	$46,945
				29706	Rich Department Store	$46,884
				30059	Gears and Parts Company	$36,782

Figure 6-33. *The expanded report*

The main drawback of this method is that you can only expand one group at a time. That's a lot of clicking to display all the details. A better solution would be to add parameters to the report to control the visibility of the groups. To do this, follow these steps:

1. In Solution Explorer, create a copy of the Sales by Territory 2 report. Name the new report Visibility by Parameters.

161

2. Add a parameter named DisplayTerritories. The Prompt should be Display Territories and the data type should be Boolean. Set the default value to False.

3. Add another parameter named DisplayDetailRows. The Prompt should be Display Detail Rows. It should also be a Boolean with a default of False.

4. Open the Group Properties of the Details group.

5. On the Visibility page, uncheck Display can be toggled by this report item.

6. Select Show or Hide based on an expression. The Dialog should look like Figure 6-34. The TerritoryID field may still be visible as the report item, but disabled.

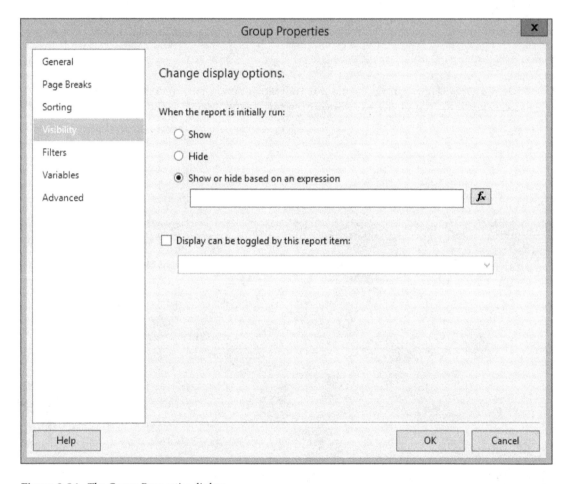

Figure 6-34. *The Group Properties dialog*

7. Click the *fx* icon to open the Expression dialog.

8. Add this expression:

```
=NOT Parameters!DisplayDetailRows.Value
```

9. Click OK to accept the expression and OK again to close the Group Properties dialog.

10. Repeat the process for the TerritoryID group. This time the expression is

    ```
    =NOT Parameters!DisplayTerritories.Value
    ```

11. Preview the report. It should look like Figure 6-35 before you change the parameters.

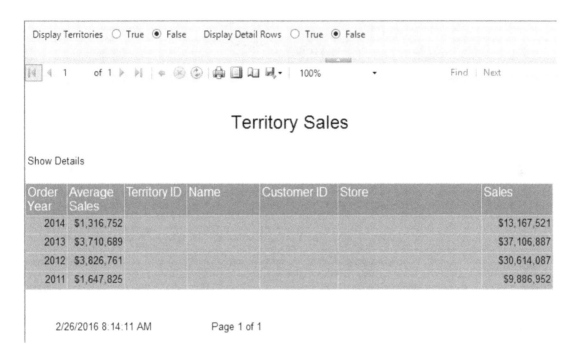

Figure 6-35. *The report with summary only*

Notice that the plus signs next to the years are gone. If you change Display Territories to True and run the report again, this time the Territories will all display. To view the details, both parameters must be set to true.

Formatting

Almost any property can be controlled by an expression. The expression may include a field from the dataset. To demonstrate, you will bold the font of the territory rows where the sales are over $2,000,000. Follow these steps:

1. Switch to design view of the Sales by Territory 2 report.

2. Select the territory row.

3. In the Properties window, locate the FontWeight property.

4. From the dropdown list, select <Expression> which opens the Expression dialog box.

5. Add this expression:

    ```
    =Iif(SUM(Fields!Sales.Value)>=2000000,"Bold","Default")
    ```

6. Click OK to save the change.

7. Preview the report and expand to see the territories. The report will look like Figure 6-36.

Territory Sales

Order Year	Average Sales	Territory ID	Name	Customer ID	Store	Sales
⊟ 2014	$1,316,752					$13,167,521
		⊞	9 Australia			$683,470
		⊞	**6 Canada**			**$2,062,731**
		⊞	3 Central			$1,077,192
		⊞	7 France			$977,725
		⊞	8 Germany			$717,890
		⊞	2 Northeast			$873,896
		⊞	**1 Northwest**			**$2,146,083**
		⊞	5 Southeast			$979,262
		⊞	**4 Southwest**			**$2,471,177**
		⊞	10 United Kingdom			$1,178,095
⊞ 2013	$3,710,689					$37,106,887
⊞ 2012	$3,826,761					$30,614,087
⊞ 2011	$1,647,825					$9,886,952

Figure 6-36. *The territories with sales at $2,000,000 or more are in bold*

The Iif function, also called inline if, takes three arguments. The first is an expression to test. If the expression returns true, then the second argument is returned. If the expression returns false, then the third value is returned. The value returned by the expression is used as the property value. In this example, the word *Bold* is returned when the sales are at least $2,000,000.

Interactive Sorting

Sorting can be controlled using a property of the cells in the header row. In my experience, it doesn't work well with collapsing sections. Follow these steps to enable interactive sorting:

1. Create a copy of Sales by Territory 2. Change the name to Interactive Sort.

2. Double-click the new report in the Solution Explorer to open it in design view.

3. Clear the visibility settings for both the TerritoryID and Details groups. Set both back to Show and uncheck Display can be toggled by this report item.

4. Preview the report to make sure that the groups and details are all visible.

5. Switch back to design view.

6. Right-click the Order Year header cell and select Text Box Properties.

7. Select the Interactive Sorting page.

8. Check Enable interactive sorting on this text box.

9. Under Choose what to sort, select Groups.

10. From the dropdown box, select OrderYear. This is the section to sort.

11. From the Sort by list, select [OrderYear]. The dialog box should look like Figure 6-37.

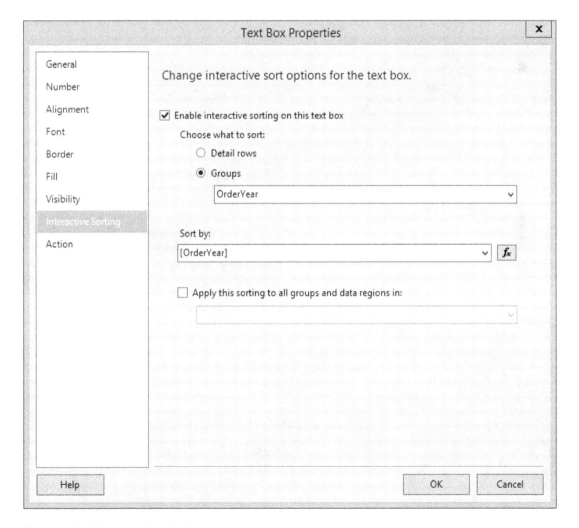

Figure 6-37. *The Interactive Sorting properties*

12. Click OK to accept the properties.

13. Set the interactive sorting property for the Territory ID header cell. The sort should apply to the TerritoryID group and sort by TerritoryID.

14. Set the interactive sorting property for the Customer ID header cell. The sort should apply to the Detail rows. Sort by CustomerID.

15. Set the interactive sorting property for the Store header cell. The sort should apply to the Detail rows. Sort by Store.

16. Preview the report. You will see arrow icons next to each sortable column as shown in Figure 6-38.

Territory Sales

Or der Ye ar	⬍	Average Sales	Territor y ID	⬍	Name	Customer ID	⬍	Store	⬍	Sales
2014		$1,316,752								$13,167,521
					9 Australia					$683,470
						29488		Nationwide Supply		$128,690
						30092		Cycle Parts and Accessories		$66,267
						29821		Bike Part Wholesalers		$59,272
						30097		Helpful Sales and Repair Service		$56,467
						29823		Popular Bike Lines		$54,683
						29628		Budget Toy Store		$46,945
						29706		Rich Department Store		$46,884
						30059		Gears and Parts Company		$36,782
						29972		Liquidation Sales		$36,585
						29595		Helmets and Cycles		$34,505
						29538		Fast Bike Works		$21,441

Figure 6-38. *The sorting icon*

Click the icons to change the sort order of the report. The first time you click, the sort will change to ascending. Click again to change to descending order. You can change the Order Year column from descending to ascending. You can switch the sort of the details between Customer ID and Store.

Creating Drill Through Reports

This chapter is full of great ways to make your reports dynamic, and the sky is the limit. There is one more feature to show you that is frequently requested. This is the ability to click a cell in a report and automatically navigate to another report containing the detail rows.

Follow these steps to create a detail report:

1. Add a new report to the project named Sales Details.

2. Add a data source to the report named AdventureWorks that points to the AdventureWorks2016 shared data source.

3. Add an embedded dataset to the report named SalesDetails. It should point to the AdventureWorks data source.

4. The query should be

```
SELECT C.CustomerID, SalesOrderID, OrderDate, TotalDue,
    T.TerritoryID, T.Name AS Territory, s.Name AS Store
FROM sales.SalesOrderHeader AS SOH
JOIN Sales.SalesTerritory AS T ON SOH.TerritoryID = T.TerritoryID
JOIN Sales.Customer AS C ON SOH.CustomerID = C.CustomerID
JOIN Sales.Store AS S ON S.BusinessEntityID = C.StoreID
WHERE YEAR(OrderDate) = @Year AND T.TerritoryID = @Territory;
```

5. Add a table control to the report.

6. Populate the table control with these fields: CustomerID, Store, SalesOrderID, OrderDate, and TotalDue,

7. Bold the header row and format the TotalDue cell as currency. The design should look like Figure 6-39.

Customer ID	Store	Sales Order	Order Date	Total Due
[CustomerID]	[Store]	[SalesOrderID]	[OrderDate]	[TotalDue]

Figure 6-39. The detail layout

8. Drag the table to the upper left and tighten up the report canvas.

9. Add a Report Page Header to the report.

10. Drag the Report Name into the header from the Built-In Fields folder.

11. Change the font size to 14 pt and expand the width of the text box.

12. Drag the Year parameter to the header.

13. Drag the Territory field from the SalesDetails dataset into the header. Since this is outside the table, it will pull the first value from the dataset. This works because the report is filtered by TerritoryID so there is only one Territory value available. The report layout should resemble Figure 6-40.

Figure 6-40. *The report layout*

At this point, it makes sense to run the report to make sure it works. When you do, just enter 2011 and 3 for the parameters. Instead of displaying parameter lists, this report will receive the values from another report. The end user will always run the report by clicking the summary report and should never need to see the parameters. Follow these steps to hide the parameters:

1. Switch back to design view of the detail report.

2. Open the properties of the Year parameter.

3. Change the Select parameter visibility property to Hidden and click OK to accept the change.

4. Repeat the process for the Territory parameter.

The next step is to add an action to a cell in a summary report. Follow these steps to add an action:

1. Create a copy of the Sales by Territory Matrix report. Name it Sales Summary.

2. Double-click the Sales Summary report to open it in design view.

3. Right-click and bring up the Text Box Properties of the cell that is at the intersection of the TerritoryID and OrderYear. Figure 6-41 shows the cell to select.

Territory ID	Name	[OrderYear]	Total
[TerritoryID]	[Territory]	[Sum(Sales)]	[Sum(Sales)]
Total		[Sum(Sales)]	[Sum(Sales)]

Figure 6-41. *The cell to select*

4. Select the Action page.

5. Select Go to report.

6. In Specify a report, select Sales Details.

7. Click Add. This adds a parameter from the Sales Details report to map to the Sales Summary report. The Name property is the parameter needed by the Sales Details report. In the Value property, you will map something from the summary report to pass through to the detail report.

8. Select the Year parameter in the Name column.

9. Select OrderYear from the Value list.

10. Click Add once more. This time, map Territory to the TerritoryID field. The dialog should look like Figure 6-42.

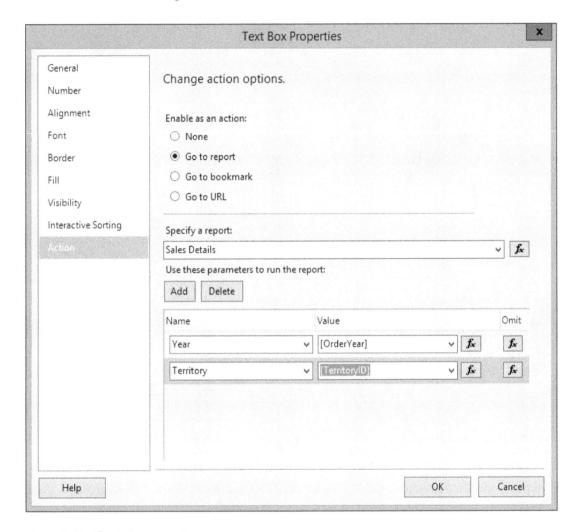

Figure 6-42. *The Action properties*

11. Click OK to accept the properties.

Now when your run the report click any of the total sales values on the Territory row. The Sales Details report will open already filtered based on the value you selected. To navigate back to the Sales Summary report, click the Back to Parent Report button.

Summary

In this chapter you saw many ways that you can make your report dynamic based on user interactivity or the data itself. You can give the user the ability to filter the data or modify properties with parameters. Text boxes and other controls can be formatted by expressions based on the data. There are no limits to what you can do.

In Chapter 7, you will learn about the visual elements that can be added to reports. These elements can be used to make a report more informative with just a glance or to create a dashboard.

CHAPTER 7

■ ■ ■

Bringing Data to Life Visually

A picture is worth a thousand words. When it comes to reports, a picture is worth a thousand numbers as well. By adding charts, gauges, and maps, a report will tell an entire story. I often joke that the higher up the organizational chart you go, the more pictures you need in the report. These pictures tell a busy executive about trends and the health of the company with just a glance.

In Chapter 4, you saw how images stored in a database can be added to a report. In Chapter 10, you will learn about an exciting new feature that also takes advantage of visual elements, Mobile Reports. In this chapter, you will add data connected visual elements to the reports and create a dashboard. The look of all the visual elements have been updated in 2016, so they will appear much different from those in previous versions.

Adding Charts and Graphs to Reports

SQL Server Reporting Services (SSRS) gives you the ability to add charts and graphs to your reports with the chart control. What is the difference between a chart and a graph? Strictly speaking, a graph displays the relationship of data over time while a chart compares categories. For example, a graph might display the sales over a year by month while a chart might compare the sales among territories for a particular year. You can see the difference in Figure 7-1.

© Kathi Kellenberger 2016
K. Kellenberger, *Beginning SQL Server Reporting Services*, DOI 10.1007/978-1-4842-1990-4_7

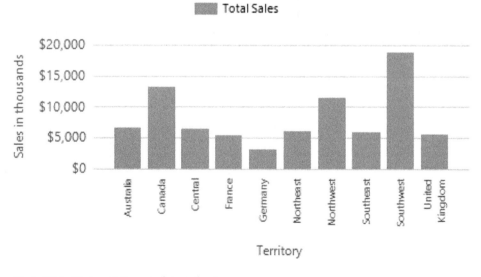

Figure 7-1. *Comparing a graph to a chart*

If you would like to see the design used for these data elements, look for a report named Chart and Graph in the Code/Download area of the Apress web site (www.Apress.com) for this book. Before diving in to create charts and graphs, take a look at Figure 7-2 so that you will understand the parts that make up one of these visual elements.

Figure 7-2. *The parts of a chart*

SSRS has a wealth of different chart types. Some of them are 3-D; however, use 3-D sparingly and with caution. Making a visual element 3-D doesn't add anything to the value, and it can distort the sizes and make understanding the data represented by the control more difficult. In this section, you will learn how to add charts and graphs to your reports. Follow these steps to get started:

1. Create a new SSRS project named Visual Reports. The solution name should be Beginning SSRS Chapter 7.

2. Add a new shared data source pointing to the AdventureWorks2016 database named AdventureWorks2016 to the project. Remember to refer back to Chapter 3 if you need to review how to add data sources and datasets.

3. Add a new shared dataset named Year to the project. It should point to the AdventureWorks shared data source. This dataset will be used in several reports, so it makes sense to be able to reuse it. Here is the query:

```
SELECT YEAR(OrderDate) AS OrderYear
FROM Sales.SalesOrderHeader
GROUP BY YEAR(OrderDate)
ORDER BY YEAR(OrderDate);
```

4. Be sure to correct the shared dataset file if the fix from Microsoft is not available. See Chapter 6 for more information.

5. Add a new report named Charts to the project.

6. Add a data source to the report named AdventureWorks pointing to the shared data source.

7. Add a new dataset to the report named Year. Select Use a shared dataset and select the Year dataset from the project. The dialog should look like that in Figure 7-3.

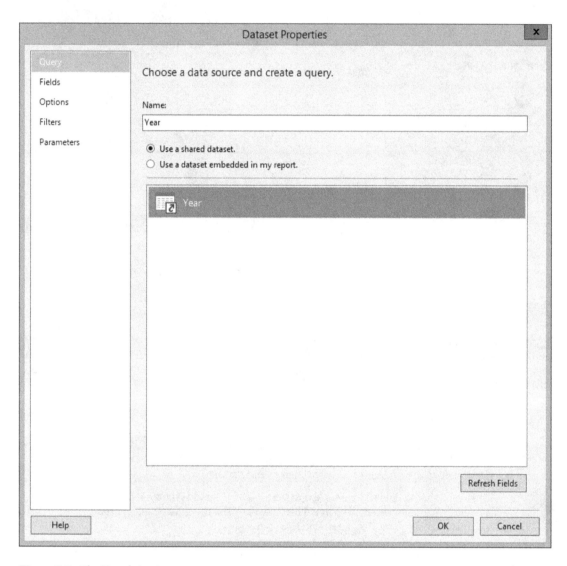

Figure 7-3. The Year dataset

8. Add an embedded dataset named Sales pointing to the AdventureWorks data source with the following query:

```
SELECT SUM(TotalDue) AS TotalSales, MONTH(OrderDate) AS OrderMonth,
    T.TerritoryID, T.Name AS TerritoryName,
    Sum(Sum(TotalDue)) OVER(PARTITION BY T.TerritoryID) AS TerritoryTotal
FROM Sales.SalesOrderHeader AS SOH
JOIN Sales.SalesTerritory AS T ON T.TerritoryID = SOH.TerritoryID
WHERE YEAR(OrderDate) = @Year
GROUP BY MONTH(OrderDate), T.TerritoryID, T.Name;
```

9. The query is filtered by @Year. The dataset will automatically create a parameter. Expand the Parameters folder and bring up the properties of the Year parameter.

10. Select the Available Values page and select Get values from a query.

11. Set Dataset to Year. Set the Value Field and Label Field to OrderYear. The Available Values page should look like Figure 7-4.

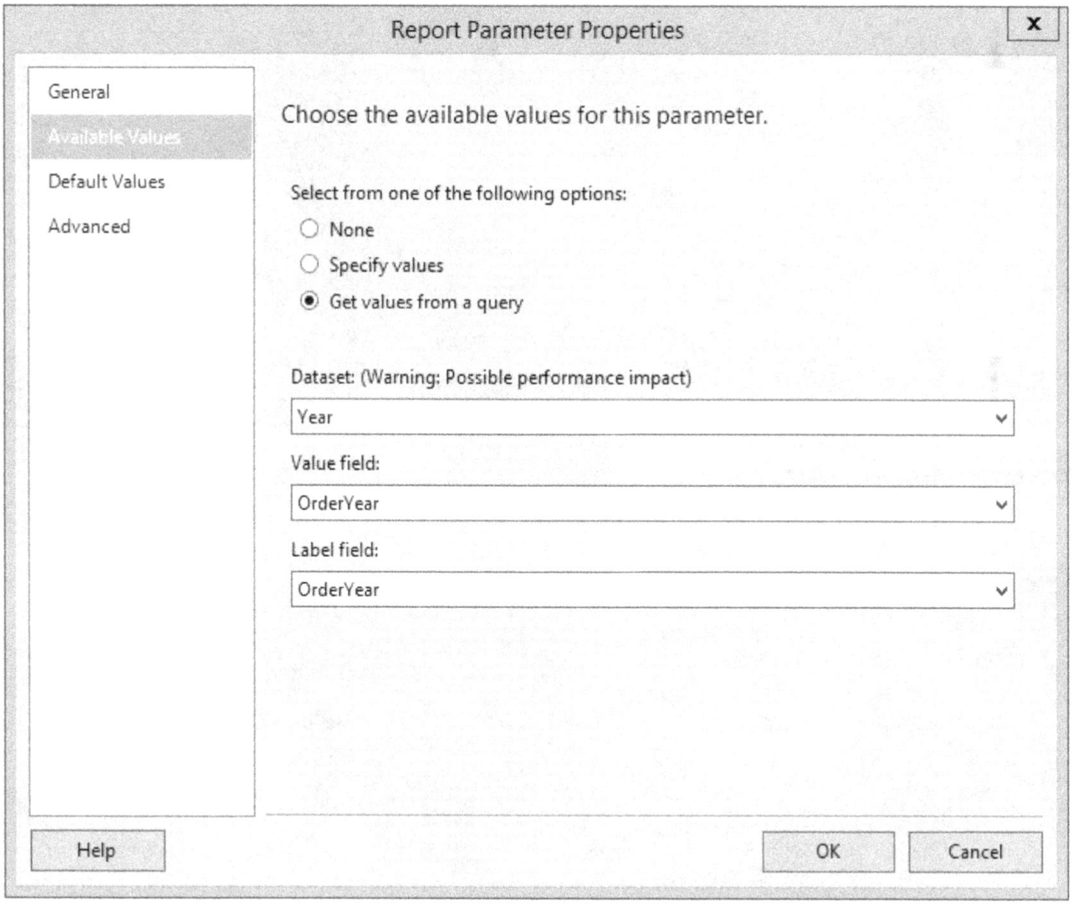

Figure 7-4. *The Available Values properties*

12. Set a default value of 2012. The Default Values page should look like Figure 7-5.

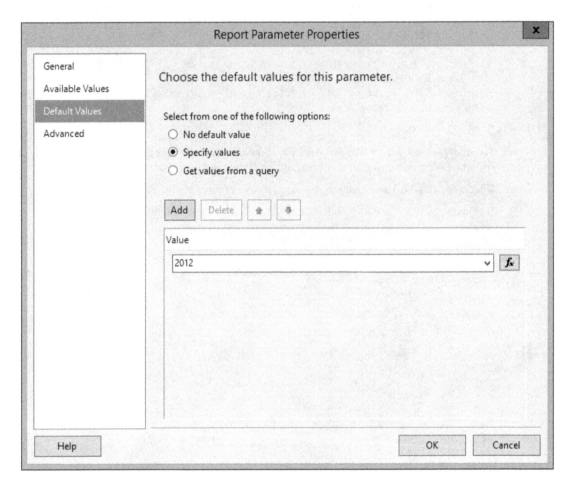

Figure 7-5. *The Default Values properties*

13. Click OK to accept the changes.

14.   The Report Data window should now look like Figure 7-6.

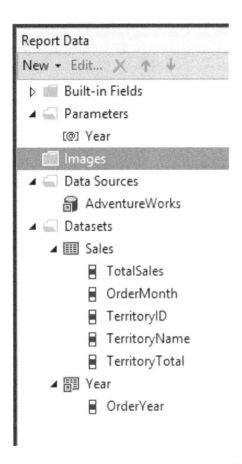

Figure 7-6. *The Report Data window after all report objects added*

Take a look at the Toolbox as shown in Figure 7-7. The Chart, Gauge, and Map are usually added as stand-alone objects in the report. The Data Bar, Sparkline, and Indicator are usually added to a cell of Tablix. In this section, you will learn about the chart control.

Figure 7-7. *The Toolbox*

Follow these steps to add a chart to the report:

1. In design view, drag a chart to the design canvas.

2. This opens the Select Chart Type dialog shown in Figure 7-8 where you can select one of many types of charts.

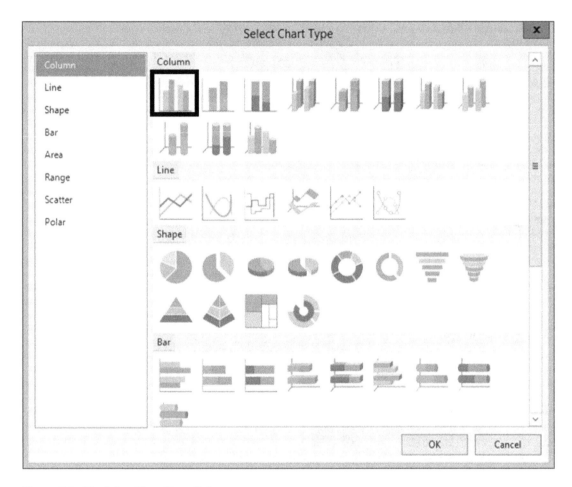

Figure 7-8. *The Select Chart Type dialog*

3. The default type is the Column chart. Click OK to add it to the report.

4. Before making any modifications, the chart will look like Figure 7-9.

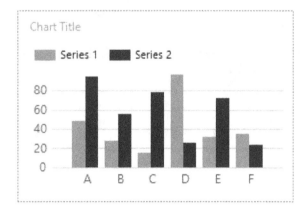

Figure 7-9. *The chart control*

5. To connect the data to the chart, double-click the chart. This opens the Chart Data window as shown in Figure 7-10.

Figure 7-10. *The Chart Data window*

6. The Chart Data window is divided into three sections. The Σ Values window holds the field you want to measure, represented by the height of the bar. Click the plus sign and navigate to the TotalSales field from AdventureWorks ➤ Sales to add it. It will automatically sum.

7. The Category Groups section holds the values used for the horizontal axis. Change from the default of Details to TerritoryName. If left as the default, it will display a bar for each row in the dataset. In our case, we want one bar per territory.

8. Series Groups allows you to break down the category into multiple smaller items. You will learn about this in a later example. For now, the Chart Data window should look like Figure 7-11.

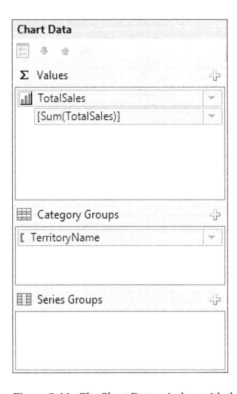

Figure 7-11. *The Chart Data window with the properties filled in*

9. Increase the size of the chart so that it is about five inches (13 cm) wide by four inches (10 cm) tall. To assist with resizing the chart, you can add the ruler by right-clicking the report and selecting View ➤ Ruler.

10. Preview the report. The chart should look like Figure 7-12.

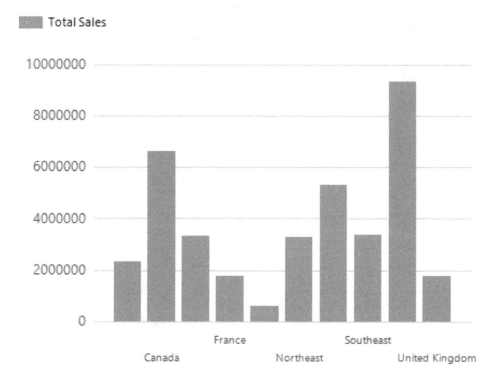

Figure 7-12. *The chart before formatting*

At this point, you can see that the bars are different heights, and you can see some of the territory names along the horizontal axis. Obviously, this chart needs quite a bit of work, and, fortunately, it is highly customizable. Each section of the chart has its own property window which you can use to control format and position. Following is a list of items that need to be corrected:

- Display all territory names.

- Add the two axis titles.

- Change the Chart Title to Total Sales by Territory along with the year.

- Remove the legend.

- Format the vertical axis.

- Add a tooltip with the exact amount for each territory.

Follow these steps to format the chart:

1. Switch back to design view.

2. Right-click the horizontal axis and select Horizontal Axis Properties.

3. This brings up the Horizontal Axis Properties dialog box. On the Axis Options page, change the Interval property to 1 and click OK. This will cause all territories to display on the chart.

4. Right-click the horizontal axis and select Show Axis Title.

5. Right-click the horizontal Axis Title and select Axis Title Properties.

6. This brings up the Axis Title Properties dialog. Change the Text Title to Territory and click OK.

7. Right-click the vertical axis and select Show Axis Title.

8. In addition to changing the axis title by bringing up the Axis Title Properties dialog, you can also change an axis title by clicking into it and typing. Change the vertical Axis Title to Sales in thousands.

9. Right-click the Chart Title and bring up the properties.

10. Next to the Title Text, click *fx* to open the Expressions dialog.

11. To add the year parameter to the title, the expression should be

    ```
    ="Total Sales by Territory for " & Parameters!Year.Value
    ```

12. Click OK twice to accept the properties.

13. Select the chart legend, Total Sales, and click the Delete key. In this case, the chart legend doesn't add anything to the report.

14. Right-click the vertical axis and select Vertical Axis Properties.

15. On the Number page of the Vertical Axis Properties dialog, select Currency under Category.

16. Change Decimal places to 0.

17. Check Use 1000 separator (,) and check Show values in Thousands. The Number page should look like Figure 7-13.

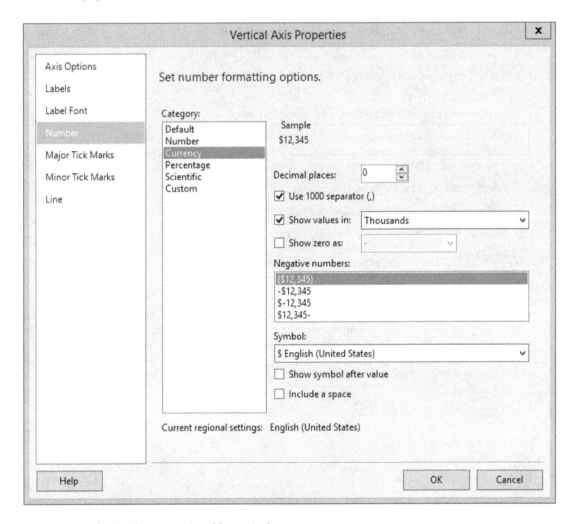

Figure 7-13. The Number properties of the vertical axis

18. Click OK to accept the properties.

19. Right-click one of the bars and select Series Properties.

20. On the Series Data page of the Series Properties dialog, click the *fx* symbol next to Tooltip.

21. The expression should be

    ```
    =FormatCurrency(Sum(Fields!TotalSales.Value),0)
    ```

22. Click OK twice to accept the change.

Now when you preview the report, the chart should resemble Figure 7-14. Be sure to hold the cursor over one or more of the bars to see the tooltip.

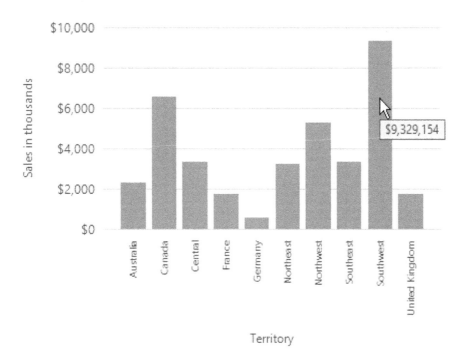

Figure 7-14. *The chart with formatting*

The report looks pretty good, but you can do more. What about sorting the bars by the total sales instead of alphabetically? To do this, follow these steps:

1. Switch to design view and double-click the chart to open the Chart Data window.

2. Click the down arrow next to TerritoryName and select Category Group Properties as shown in Figure 7-15.

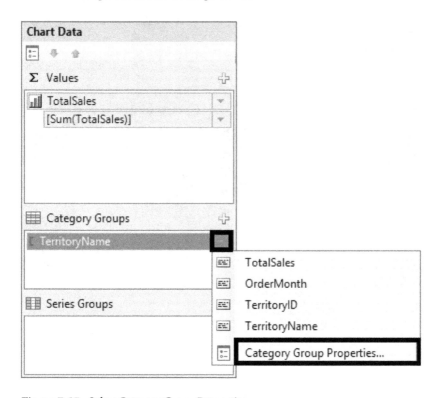

Figure 7-15. *Select Category Group Properties*

3. Select the Sorting page and change the Sort by property from TerritoryName to TerritoryTotal. This field has been pre-aggregated as a sum for each territory. You cannot use the Sum(TotalDue) expression to sort.

4. Click OK and preview the report. The report should look like Figure 7-16.

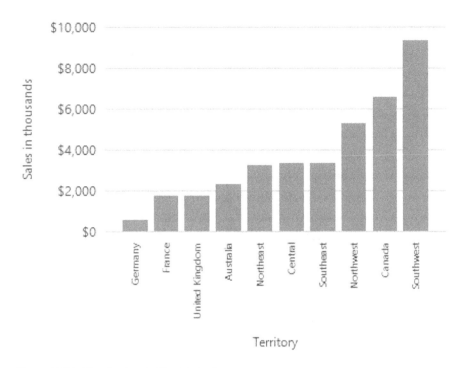

Figure 7-16. *The chart sorted by total sales*

This chart type is probably the best way to display this information, but there are several other types you could use. A pie chart could be used, but it would be difficult for the user to understand because of the high number of categories. Funnel or pyramid charts may work as well, but you could also choose the tree map control. This type of chart is new with SSRS 2016.

Follow these steps to learn how to use the new tree map chart:

1. Switch back to design view.

2. Increase the length of the report canvas to make room for the next chart.

3. Add a chart to the report.

4. On the Select Chart Type dialog, select the Tree Map as shown in Figure 7-17 and click OK.

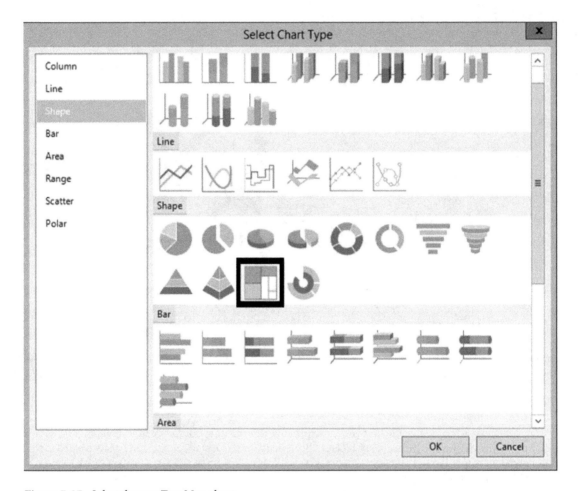

Figure 7-17. *Select the new Tree Map chart*

5. Increase the size of the chart.

6. Open the Chart Data window by double-clicking the chart.

7. Add TotalSales to the ∑ Values section.

8. Set the Category Groups values to TerritoryName. This will create a section for each territory, but they will be the same color.

9. To make each territory a different color, also set the Series Groups to TerritoryName.

10. Preview the report. At this point, the chart will look like Figure 7-18.

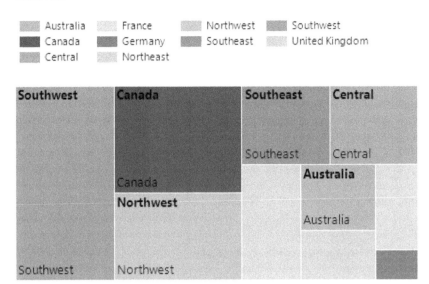

Figure 7-18. *The Tree Map chart before formatting*

Notice that some of the cells have the territory displayed twice, but some do not have anything displayed. To get around this, use a tooltip to display the information. Follow these steps to add the tooltip and complete the formatting:

1. Switch back to design view.

2. Right-click a series cell and click Show Data Labels as shown in Figure 7-19. This actually switches the labels at the bottom to the total sales.

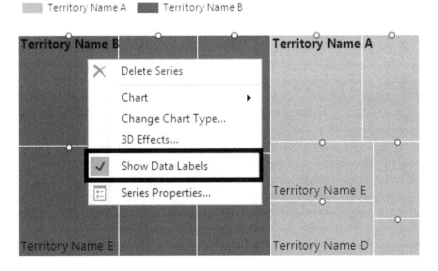

Figure 7-19. *Show Data Labels*

3. Right-click one of the numbers and select Series Label Properties.

4. Select the Number page of the Series Label Properties dialog and format as Currency with no decimal places and use the 1000 separator.

5. Click OK to accept the change.

6. Right-click a series cell and select Series Properties.

7. Click *fx* next to the ToolTip property and use the following expression:

```
=Fields!TerritoryName.Value & " " &
FormatCurrency(Sum(Fields!TotalSales.Value),0)
```

8. Click OK twice to dismiss the dialogs.

9. Change the Chart Title to the following expression:

```
="Sales by Territory for " & Parameters!Year.Value
```

10. Preview the report. The chart should look similar to Figure 7-20.

Figure 7-20. *The formatted tree map chart*

The Line chart type is also interesting. With it you can create a graph that compares multiple categories over time. In this case, you will create a graph that displays the sales for the territories over each month for one year. Follow these steps to create the graph:

1. Switch back to design view.

2. Expand the size of the report canvas.

3. Add a new chart to the report as shown in Figure 7-21.

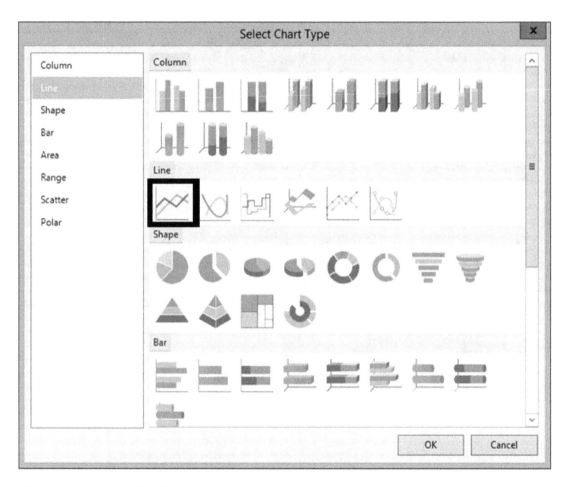

Figure 7-21. *Select the Line chart*

4. Increase the size of the chart.

5. Set the Chart Data properties as shown in Figure 7-22. The Series Groups property will split the line into individual territories.

Figure 7-22. *The Chart Data properties*

6. Change the Interval property of the horizontal axis to 1 so that all months will display.

7. To change the months in the horizontal axis to the month name instead of number, click the down arrow in the Chart Data window next to OrderMonth and select Category Group Properties as shown in Figure 7-23.

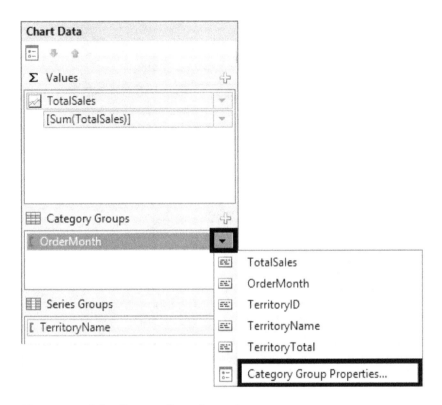

Figure 7-23. *Select Category Group Properties*

8. Click the *fx* icon next to Label and add this expression:

    ```
    =MonthName(Fields!OrderMonth.Value)
    ```

9. Click OK twice to dismiss both dialog boxes.

10. Select one of the series lines and select Series Properties.

11. Change the tooltip property to this expression:

    ```
    =Fields!TerritoryName.Value & " " & FormatCurrency(Fields!TotalSales.Value,0)
    ```

12. Click OK to dismiss the Expressions dialog and then select the Markers page.

13. Change the Marker Type to Circle and click OK.

14. Right-click the legend and bring up the properties.

15. Change the Legend position to the top as shown in Figure 7-24.

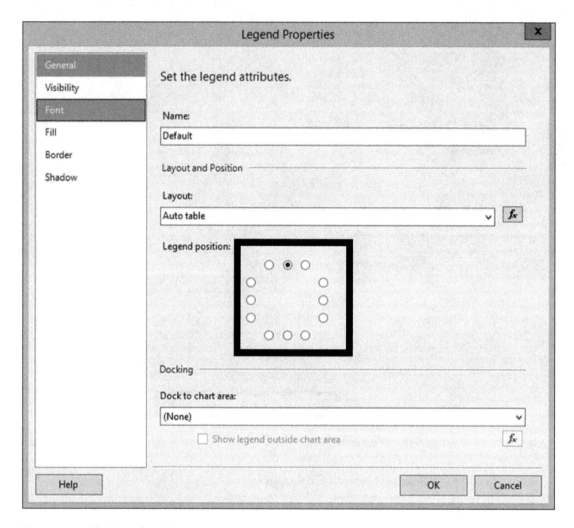

Figure 7-24. *The Legend position*

16. Click OK to accept the change.

17. Change the Chart Title to

    ="Sales by Month for " & Parameters!Year.Value

18. Bring up the Vertical Axis properties and format the vertical axis so that it displays currency with no decimal places and with a thousands separator.

19. Only show values in thousands and click OK to save the changes.

20. Right-click the vertical axis and select Show Axis Title. The title should say Sales in Thousands.

21. Set the Interval property of the horizontal axis to 1 so that all months display.

22. Preview the report. It should look similar to Figure 7-25.

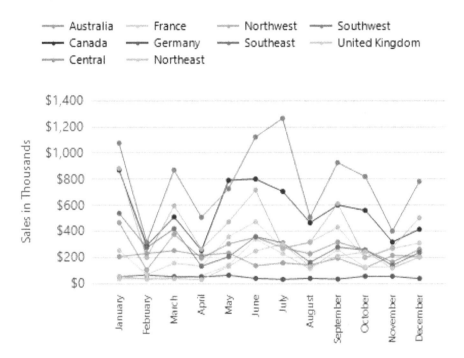

Figure 7-25. *The sales graph*

Adding Gauges to Reports

Charts and graphs compare values over categories, time, or both. A gauge can be used to show if a single goal was met. Meeting a sales quota is a good example. Gauges look like thermometers and dials, and they are more complex to configure than charts.

Before diving in to the details, take a look at the different parts that make up a gauge as shown in Figure 7-26.

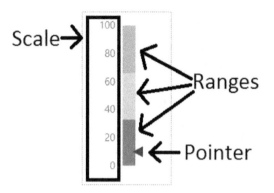

Figure 7-26. *The parts of a gauge*

The scale represents the range of values you expect. It can be in terms of a numeric value such as dollars or a percent. The top value of the scale could be the goal or quota. However, if it is possible to exceed the goal, you may want to make the top value higher than the goal, such as the goal plus 25%.

The range highlights certain areas on the scale. For example, you may want to highlight the goal. You can customize the range with color, and one way might be to color the range from red to green. It's possible to add multiple ranges to have even more granular control over the color.

The pointer represents the value achieved. At a glance, you can tell if the goal was met by how close it lands to the target. Several types of gauges are available, including those with two dials and those with logarithmic scales. To learn how to work with gauges, follow these steps:

1. Add a new report to the project named Gauges.

2. Set up a data source pointing to AdventureWorks2016 named AdventureWorks.

3. Add a dataset named Year with pointing to the Year shared dataset.

4. There is a table with quotas in the AdventureWorks database, but the values do not make much sense compared to the sales. Instead, create a dataset named SalesQuota with this query containing hard-coded values:

```
SELECT * FROM
    (VALUES(2011,1000, 899),
        (2012,1000,1010),
        (2013,1200,1100),
        (2014,1200,1220))
    AS Quota ([Year],[Target],Sales)
WHERE [Year] = @Year;
```

5. A Year parameter will automatically be created. Change the Available Values to Get values from a query. The Dataset is Year. The Value field and Label field should be set to OrderYear.

6. Add a gauge to the report. Select the Radial graph as shown in Figure 7-27.

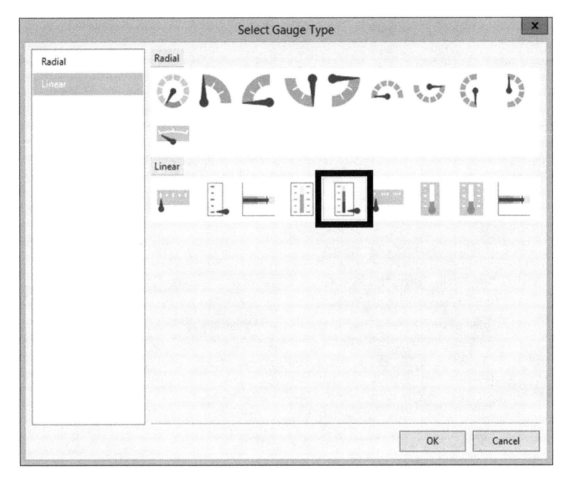

Figure 7-27. *Add a Radial gauge to the report*

7. When you click the gauge, the Gauge Data window opens. Under LinearPointer1, change Unspecified to Sales. It will automatically sum as shown in Figure 7-28.

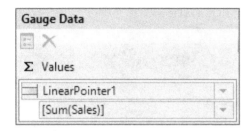

Figure 7-28. *The Gauge Data properties*

8. Right-click inside the gauge. Select Gauge Panel ➤ Scale Properties.

9. By default, the scale goes from 0 to 100. To change it to match the values in the data, click the *fx* symbol for the Maximum property on the General page.

10. Change the expression to

    ```
    =Fields!Target.Value * 1.25
    ```

11. Switch to the Number page and format as Currency with no decimal places.

12. Click OK to accept the properties.

13. Bring up the Range Properties of the top range by right-clicking the gauge and selecting Gauge Panel ➤ Range (LinearRange3) Properties.

14. On the General page, change the Start range at scale value property to the following expression:

    ```
    =Fields!Target.Value * .95
    ```

15. Change the End range at scale value to this expression:

    ```
    =Fields!Target.Value * 1.25
    ```

16. Click OK to accept the properties.

17. Bring up the Range Properties of the middle range, LinearRange2.

18. Set the Start range at scale value to

    ```
    =Fields!Target.Value * .75
    ```

19. Set the End Range at Scale Value to

    ```
    =Fields!Target.Value * .95
    ```

20. Click OK.

21.   Bring up the Range Properties of the bottom range, LinearRange1.

22.   Set the End range at scale value to

```
=Fields!Target.Value * .75
```

23.   Click OK.

24.   Right-click the scale and select Scale Properties.

25.   Select the Labels page.

26.   Uncheck Show labels at end of scale.

27.   Click OK.

Now when you preview the report and select 2014 for the Year parameter, the gauge will look like Figure 7-29.

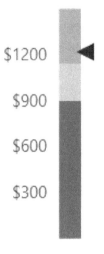

Figure 7-29. *The configured gauge*

While you will generally see gauges as a stand-alone section of a report, you can also add gauges to a table cell. Follow these steps to learn how to do it:

1.   Switch to design view.

2.   To display all years in one report, add a new dataset named SalesAllYears with the following query:

```
SELECT * FROM
    (VALUES(2011,1000, 899),
        (2012,1000,1010),
        (2013,1200,1100),
        (2014,1200,1220))
    AS Quota ([Year],[Target],Sales);
```

3. Add a table to the report.

4. Add the fields Year, Target, and Sales to the table from the SalesAllYears dataset.

5. Add a new column to the right.

6. Drag a gauge into the new data cell.

7. Select the Linear Horizontal guage as shown in Figure 7-30.

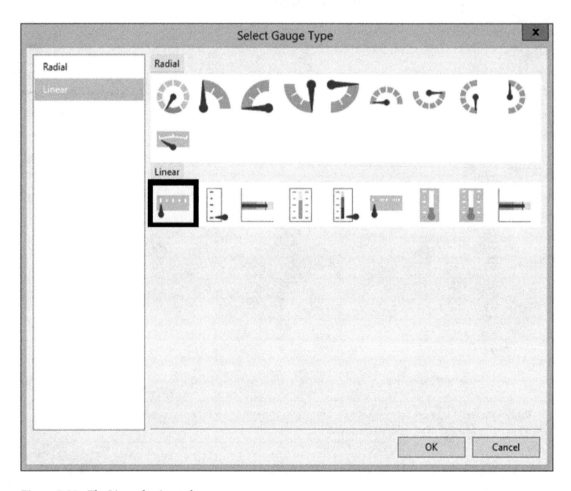

Figure 7-30. The Linear horizontal gauge

8. Expand the height and width of the cell so that the table design looks like Figure 7-31.

Year	Target	Sales	
[Year]	[Target]	[Sales]	

Figure 7-31. *The table design*

9. Double-click the gauge to bring up the Guage Data window. Change the value under LinearPointer1 to Sales. It will automatically sum.

10. Bring up the Scale Properties and change the Maximum to the following expression:

    =Fields!Target.Value * 1.25

11. Change the Interval property to 200.

12. On the Number page, change the Category to Currency.

13. Change to 0 decimal places and check Use 1000 separator (,).

14. On the Labels page, uncheck Show labels at the end of scale.

15. Click OK to accept the changes.

16. Right-click the gauge and select Add Range.

17. Bring up the Range properties.

18. Change both Start range at scale value and End range at scale value to Sum(Target).

19. On the Border page, change the Line Width to 5 pt.

20. Click OK to accept the changes.

21. Select the cell that holds the gauge.

22. In the Properties window, change the BorderStyle Default property to Solid and BorderColor property to LightGray.

When you preview the report, the table should look like Figure 7-32.

Figure 7-32. *The table with an embedded gauge*

Adding Data Bars, Sparklines, and Indicators to Tables

You saw that you can add gauges to table cells in the last section. Three visual controls, data bars, sparklines and indicators, are meant to be used within cells. Indicators are simplified gauges. Data bars and sparklines are simplified charts and graphs. To learn how to add an indicator, follow these steps:

1. Add a new report named SmallControls to the project.

2. Add the AdventureWorks data source pointing to the shared AdventureWorks2016 dataset.

3. Add an embedded dataset named Sales with the following query:

    ```
    SELECT YEAR(OrderDate) AS OrderYear, MONTH(OrderDate) AS OrderMonth,
        SUM(TotalDue) AS Sales
    FROM Sales.SalesOrderHeader
    GROUP BY YEAR(OrderDate), MONTH(OrderDate);
    ```

4. Add a table to the report.

5. Add the Sales field to the detail row.

6. Right-click the detail row and select Add Group ➤ Parent Group.

7. In the Tablix Group dialog, select OrderYear for the Group by property.

8. Check Add group header. The dialog should look like Figure 7-33.

9. Click OK.

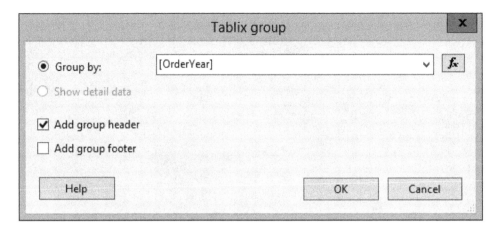

Figure 7-33. *The OrderYear group*

10. Delete the detail row by right-clicking the row and selecting Delete rows.

11. Click OK when asked if you want to delete the row and group.

12. In the data cell under Sales, select Sales. It will automatically sum.

13. Drag an Indicator from the Toolbox to the third data cell in the row.

14. Select the default Directional style and click OK.

15. Delete the empty column.

16. Click the cell with the Indicator. You'll see GaugePanel1 in the Properties window. Change the BorderStyle Default property to Solid.

17. Double-click the cell holding the indicator. Change the Gauge Data value to Sales. It will automatically sum as shown in Figure 7-34.

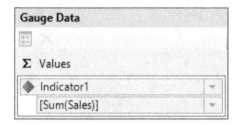

Figure 7-34. *The Gauge Data properties*

When you run the report, it will look like Figure 7-35.

Order Year	Sales	
2011	14155699.5250	↓
2012	37675700.3120	↑
2013	48965887.9632	↑
2014	22419498.3157	↓

Figure 7-35. *The report with the indicator*

By default, the indicator is set to display the three possible icons based on the percentage of sales. In design view, right-click the cell and select Indicator Properties. You can change the behavior of the indicator on the Value and States page as shown in Figure 7-36.

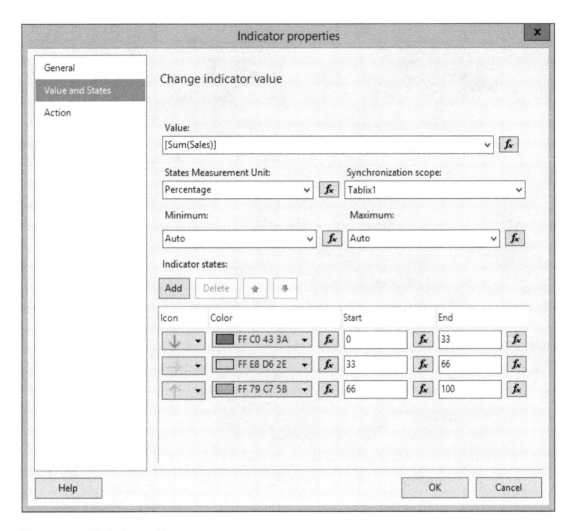

Figure 7-36. *The Value and States page*

Change the Start and End values to match Figure 7-37.

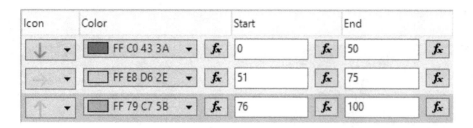

Figure 7-37. *The new indicator properties*

Click OK to accept the changes. Now when you run the report, all sales indicators at the 50th percentile will be the red down arrow as shown in Figure 7-38.

Order Year	Sales	
2011	14155699.5250	↓
2012	37675700.3120	→
2013	48965887.9632	↑
2014	22419498.3157	↓

Figure 7-38. *The results after the indicator change*

This report displays the data at the year level. You can use a sparkline to display the sales over the months. Follow these steps to learn how:

1. Switch to design view.

2. Add a new column to the right of the indicator.

3. Drag a sparkline control to the new data cell.

4. Select the Line with Markers sparkline type as shown in Figure 7-39 and click OK.

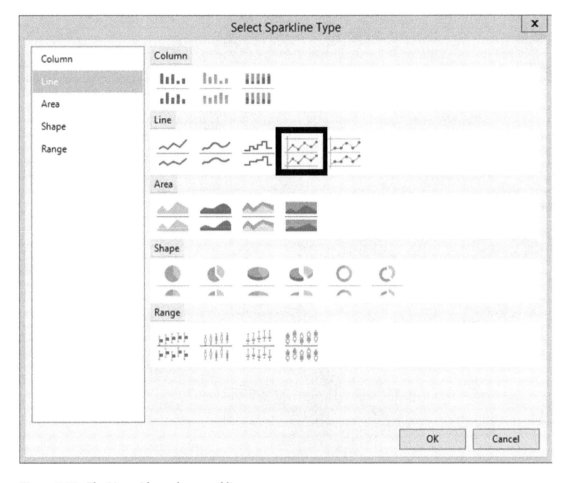

Figure 7-39. The Line with markers sparkline

5.   Double the width of the table cell containing the sparkline.

6.   Double-click the cell to bring up the Chart Data properties.

7.   Change the ∑ Value to Sales and the Category Groups to OrderMonth.

8.   Click the sparkline until the markers light up.

9.   Right-click the sparkline and select Series Properties.

10. Change the ToolTip property to the following expression and click OK:

```
=MonthName(Fields!OrderMonth.Value) & " "
    & FormatCurrency(Fields!Sales.Value,0)
```

11. Select the cell with the sparkline. In the Properties window, change the BorderStyle Default property to Solid.

Now, when you run the report, you can run your cursor over the line to see the month name and the total sales for that month. Figure 7-40 shows how the report should look:

Figure 7-40. *The report with a sparkline*

A data bar can be added and configured just like the sparkline. The difference is that there will be separate bars for each data point instead of a line. Follow these steps to add the data bar:

1. Switch to design view.

2. Add a new column to the table.

3. Drag a data bar to the new cell.

4. Select the Data Column type as shown in Figure 7-41.

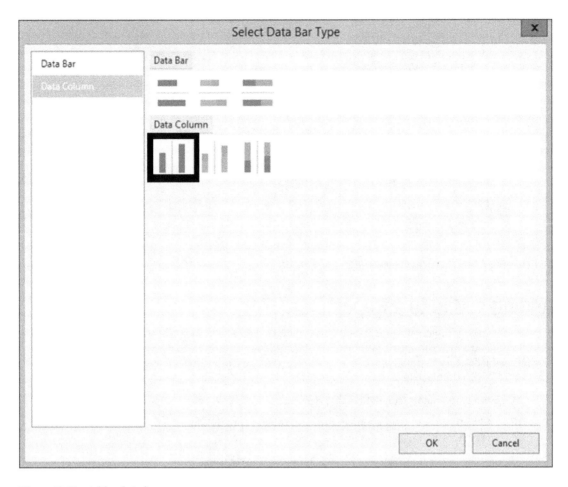

Figure 7-41. *Add a data bar*

5. Expand the width of the cell.

6. In the Chart Data window, select Sales under the ∑ Value property.

7. Add OrderMonth to the Category Groups section.

8. In this case, you will sort by value instead of the month. Click the arrow next to OrderMonth in the Category Groups section.

9. Select Category Group Properties.

10. On the Sorting page, replace OrderMonth with Sales.

11. Click OK.

12. Click the cell. In the Properties window, change the BorderStyle Default property to Solid.

13. Click one of the data bars until a small circle appears above each of them.

14. Right-click and bring up Series Properties.

15. Add the following expression to the Tooltip property:

```
=MonthName(Fields!OrderMonth.Value) & " "
    & FormatCurrency(Fields!Sales.Value,0)
```

When previewing the report, it should look like Figure 7-42.

Order Year	Sales			
2011	14155699.5250			
2012	37675700.3120			
2013	48965887.9632			May $3,452,924
2014	22419498.3157			

Figure 7-42. *The report with the data bars*

Adding a Map to a Report

Arguably the most intriguing feature of SSRS is the ability to add maps to reports. In this section, you will learn how to add a map connected to geographic data from AdventureWorks. This just touches the surface of what can be done with maps. In fact, an entire book could be written on this topic. Follow these steps to learn how to add a simple map to a report:

1. Add a new report named Map to the project.

2. Add a data source named AdventureWorks pointing to the shared AdventureWorks2016 data source.

3. Add a dataset named Year pointing to the shared Year dataset.

4. Add an embedded dataset named MapData with the following query:

```
DECLARE @Year INT = 2013;
SELECT SUM(TotalDue) AS Sales, vS.CountryRegionName, vS.StateProvinceName
FROM Sales.SalesOrderHeader AS SOH
JOIN Person.BusinessEntityAddress AS BEA
    ON BEA.BusinessEntityID = SOH.CustomerID
JOIN Person.Address AS A ON BEA.AddressID = A.AddressID
JOIN Person.vStateProvinceCountryRegion AS vS
    ON A.StateProvinceID = vS.StateProvinceID
WHERE CountryRegionName = 'United States'
    AND YEAR(OrderDate) = @Year
GROUP BY A.City, vS.CountryRegionName, vS.StateProvinceName;
```

5. The Year parameter will be added automatically. Connect the available values to the Year dataset. At this point, the MapData dataset is hard-coded for 2013, which is needed to get the map wizard to work. Once the map is done, you will make the report dynamic.

6. Add a map control to the report which kicks off a wizard.

7. On the Choose a source of spatial data page, select Map gallery and choose USA by State as shown in Figure 7-43.

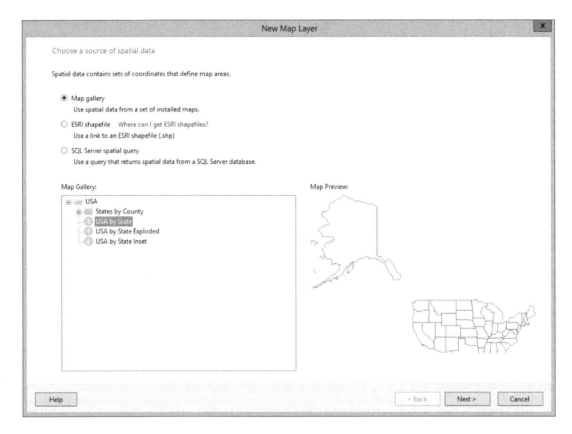

Figure 7-43. *Select the source of map data*

8. Click Next.

9. On the Choose Spatial Data and Map View Options page, you can modify the resolution, position, and zoom level of the map. Make the map slightly smaller by dragging down the pointer on the left. Adjust the position by grabbing the map and moving it so that Alaska, Hawaii, and the continental United States are visible. Click Next.

10. On the Choose Map Visualization page, select the Color Analytical Map and click Next.

11. On the Choose the Analytical Dataset, select MapData and click Next.

12. The Specify the Match Fields for Spatial and Analytical Data page is where you connect the map properties to fields in the data. Check the box next to STATENAME.

13. In the Analytical Dataset Fields dropdown box, select StateProvinceName. You can verify that you made the correct choices by comparing the highlighted columns in the two bottom sections as shown in Figure 7-44.

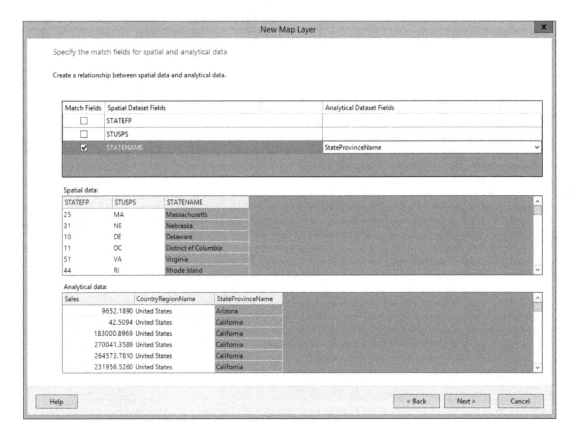

Figure 7-44. *Map data connected to the dataset column*

14. Click Next.

15. On the Choose color theme and data visualization page select Sum(Sales) from the Field to Visualize list.

16. In the Color Rule list, select Red-Yellow-Green.

17. The page should look like Figure 7-45.

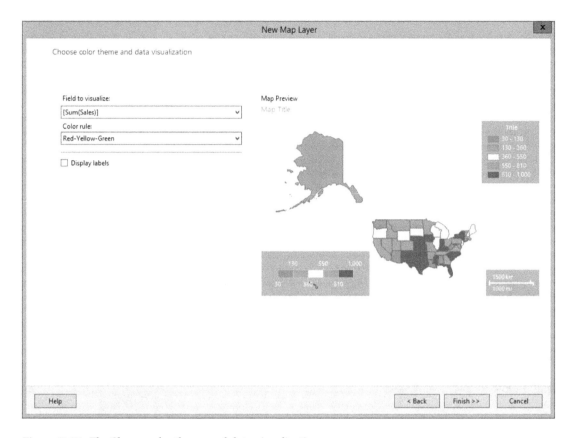

Figure 7-45. *The Choose color theme and data visualization page*

18. Click Finish.

19. Open the MapData dataset properties.

20. Remove this text from the following query:

```
DECLARE @Year INT = 2013;
```

21. Click OK to accept the change. In order to connect the data to the map, a value for the parameter had to be provided. Now that the map is complete, this line can be removed.

When you preview the report and select 2012, you should see the populated map as shown in Figure 7-46.

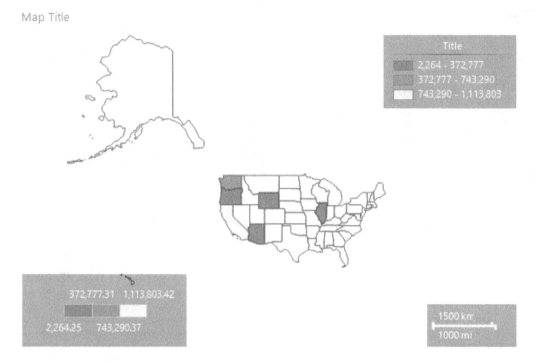

Figure 7-46. *The populated map*

To add a tooltip to the map follow these steps:

1. Switch back to design view.

2. Click the map to open the Map Layers window.

3. Click the down arrow in the PolygonLayer and select Polygon Properties as shown in Figure 7-47.

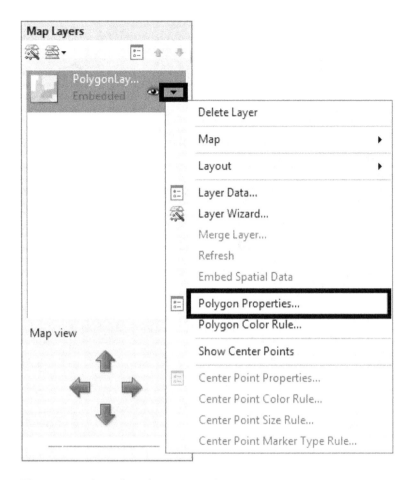

Figure 7-47. *Open the polygon properties*

4. Change the Tooltip property to the following expression:

    =Fields!StateProvinceName.Value & " " & FormatCurrency(Fields!Sales.Value,0)

5. Change the Map Title to the following expression:

    ="US Sales by State " & Parameters!Year.Value

6. Right-click the map legend and bring up the properties. Change the Legend Position to the bottom of the map.

215

7.   Check Show legend outside the viewport as shown in Figure 7-48.

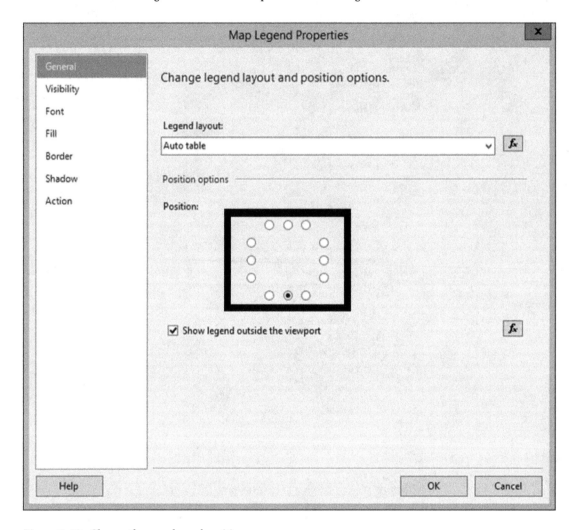

Figure 7-48. *Change the map legend position*

8.   Click OK.

9.   Change the map legend title from Title to Sales.

10.   Remove the Color Scale.

11.   Remove the Distance Scale.

When you run the report and select 2012, the report should look like Figure 7-49.

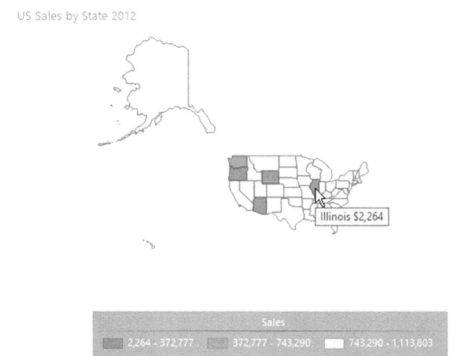

US Sales by State 2012

Illinois $2,264

Sales

2,264 - 372,777   372,777 - 743,290   743,290 - 1,113,803

Figure 7-49. *The final map report*

Building a Dashboard

Dashboards allow busy executives to see the current status of the business with just a glance. They can see trends over time and how key performance indicators are being met. In this section, you will combine several of the visualizations you created into one report.

To get started, follow these steps:

1. Add a new report named Dashboard.

2. Open the Report Properties from the Report menu.

3. Change the layout of the report to Landscape.

4. Change the margins so that each margin is 0.25 inches or 0.625 cm as shown in Figure 7-50.

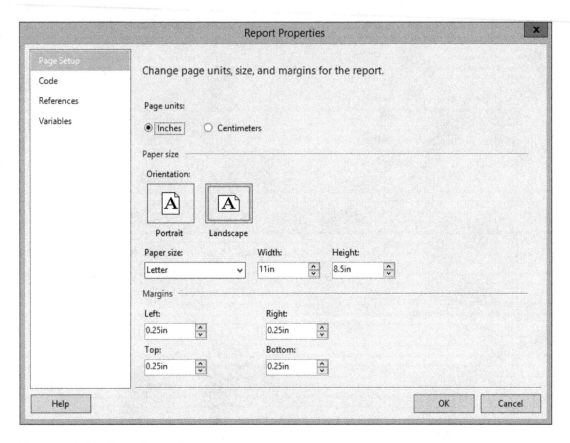

Figure 7-50. *The Report Properties page*

5. Click OK to save the changes.

6. Add a data source to the report named AdventureWorks pointing to the
    AdventureWorks2016 shared data source.

7. Add a dataset named Year pointing to the shared Year dataset.

8. Add a parameter named Year of data type integer. The General page should look like Figure 7-51.

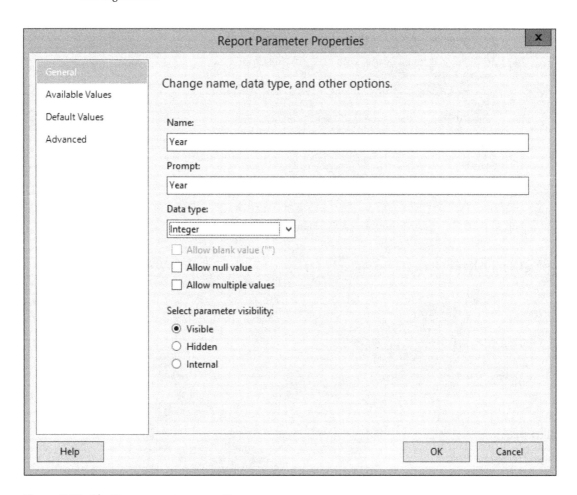

Figure 7-51. *The Year parameter properties*

9.   Connect the available values to the Year dataset.

10.   Add a default to the parameter with the value 2014.

11.   Click OK.

12.   Create a new embedded dataset named Sales pointing to AdventureWorks with the following query:

```
SELECT SUM(TotalDue) AS TotalSales, MONTH(OrderDate) AS OrderMonth,
    T.TerritoryID, T.Name AS TerritoryName,
    Sum(Sum(TotalDue)) OVER(PARTITION BY T.TerritoryID) AS TerritoryTotal
FROM Sales.SalesOrderHeader AS SOH
```

```
JOIN Sales.SalesTerritory AS T ON T.TerritoryID = SOH.TerritoryID
WHERE YEAR(OrderDate) = @Year
GROUP BY MONTH(OrderDate), T.TerritoryID, T.Name;
```

13. Add a page header to the report.

14. Add a text box to the header with the following expression:

    ```
    ="Sales Dashboard for " & Parameters!Year.Value
    ```

15. Increase the font size of the text box to 16 pt.

16. Expand the width of the text box.

17. Open the Charts report in design view.

18. Copy the bar chart from the report using the CTRL+C shortcut and paste it into the Dashboard report.

When you run the report, it should look like Figure 7-52.

Sales Dashboard for 2014

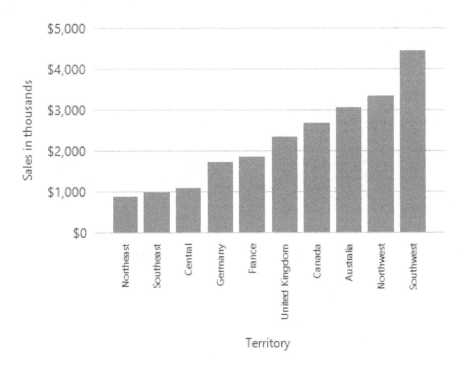

Figure 7-52. The dashboard with the first visualizations

You can continue adding embedded visualizations to the dashboard, but you can also add subreports that contain tables or visualizations. Follow these steps to add subreports:

1. Switch to design view of the Map report.

2. Drag the map to the upper left corner of the report.

3. Drag in the margins of the report as far as possible.

4. Save the report.

5. In design view of the Dashboard report, add a subreport control to the right of the bar chart.

6. Adjust the size of the subreport so that it is about the same size as the bar chart.

7. Right-click the subreport and select Subreport Properties.

8. Set the Use This Report as a Subreport property to Map.

9. On the Parameters page you map any parameters needed for the subreport to values from the parent report. Click Add to add the first parameter.

10. Select Year under Name. This is the parameter required by the subreport.

11. Under Value, select the *fx* icon to open the Expression builder.

12. Change the expression to =Parameters!Year.Value and click OK twice.

13. Preview the report. You can adjust the size of the map in the Map report and the subreport as required. The report should look like Figure 7-53.

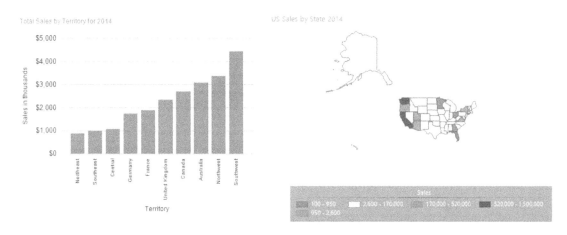

Figure 7-53. The dashboard with two visualizations

14. Open the SmallControls report in design view.

15. Drag the table to the upper left corner.

16. Drag in the report edges.

17. If you haven't done it previously, format the sales field as currency with no decimal places and with a thousands separator.

18. Save the SmallControls report.

19. Add the SmallControls report as a subreport to the Dashboard report below the bar chart. This subreport doesn't have a parameter to map.

When you view the report, it should resemble Figure 7-54.

Figure 7-54. *The completed dashboard*

Summary

SSRS provides a wide variety of visual elements that connect to datasets. These objects also have many configurable properties that can enhance the visual appeal of reports and dashboards. This chapter introduced you to all of the charts, gauges, and maps, including several that are meant to fit inside a Tablix cell.

In Chapter 8, you will learn how to publish reports to the new Reporting Services Web Portal.

PART III

Deploying Reports

■ ■ ■

CHAPTER 8

Publishing Reports

I write articles, blog posts, books, and a monthly newsletter. Until these are published, they exist on my laptop where only I can see them. The point of my writing is to make it available for others to read. My goal is to teach people all over the world about SQL Server, and to accomplish this, my works must be published.

You can create brilliant reports using SQL Server Data Tools (SSDT) on your computer, but until those reports are published, they are not useful for the people who requested them. In this chapter, you will learn how to publish the reports that you have created throughout the book.

Getting Around in the Web Portal

Throughout all the versions of SQL Server Reporting Services (SSRS), the default user interface for SSRS reports when installed in native mode was the Report Manager as shown in Figure 8-1.

Figure 8-1. The Report Manager

Starting with SSRS 2016, the Report Manager has been replaced with the new SSRS web portal. This change not only represents an upgrade but enables SSRS to play in the new mobile world with new report types intended for smartphones and tablets. If you have managed SSRS reports over the past few years, you have probably received complaints about web browser compatibility. Starting with 2016, modern web browsers are supported!

Over the past few releases, many of the new SSRS features were available only when SSRS was installed in SharePoint integrated mode. This release is a nice departure from that trend. Those SharePoint-only features, such as data-driven reports, have not been back-ported to native mode, but several new features are available in native mode only. Figure 8-2 shows the new portal.

© Kathi Kellenberger 2016
K. Kellenberger, *Beginning SQL Server Reporting Services*, DOI 10.1007/978-1-4842-1990-4_8

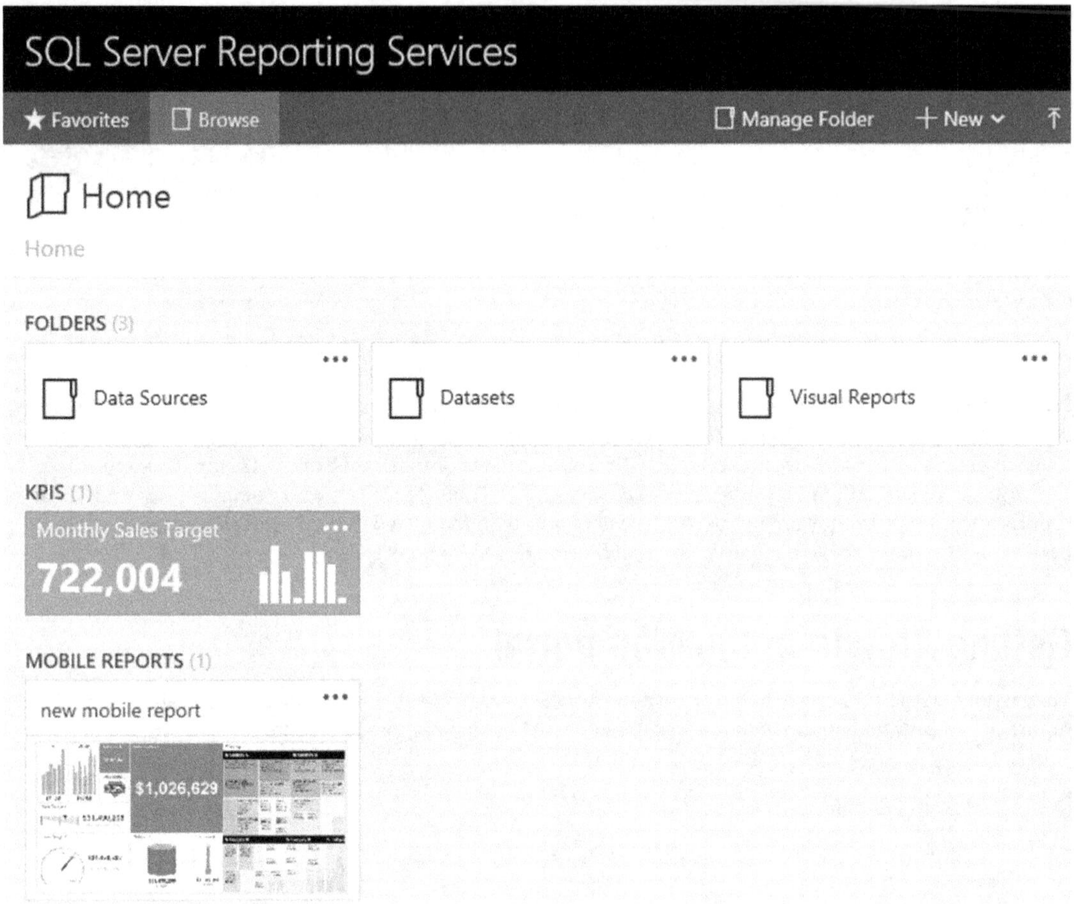

Figure 8-2. *The new web portal*

In addition to hosting the traditional paginated reports, the web portal hosts independent KPIs, or key performance indicators, and mobile reports that will run on tablets and smartphones. You will learn how to build the KPIs and mobile reports in Chapter 10.

The web portal has two pages or modes: Favorites and Browse. The Browse page looks more like the traditional view. There you can see the folders where the reports and other objects are stored as shown in Figure 8-3.

Figure 8-3. *The Browse page*

When you click a folder, such as the Reports folder, you will see the objects within that folder. Figure 8-4 shows the contents of the Reports folder.

Figure 8-4. *The Reports folder*

Inside each folder, you can store KPIs, mobile reports, and paginated reports. You will also have folders, usually in the Home folder, for data sources and the other objects that can be published. Across the top of the folder, you can see a path back to the Home folder. To run a paginated report, you just have to navigate to it and then click it.

The Favorites page is populated specifically by each user. To add a report to the Favorites page, just click the ellipsis next to the object and click Add to Favorites as shown in Figure 8-5.

Figure 8-5. *Add a report to the Favorites page*

To accomplish many of the management tasks in the web portal, you will need to start with the menu at the top right of the page as shown in Figure 8-6.

Figure 8-6. *The menu*

End users just need to know where to find the reports they want to run. They can make things more convenient for themselves by utilizing the Favorites page. SSRS developers and administrators need to understand much more. The rest of the chapter will cover the tasks that developers and administrators do to provide a convenient reporting experience.

Deploying Reports from SSDT

Reports can be deployed directly from SSDT or by uploading directly within the web portal. When deploying from SSDT, any required folders are created as well. To get started, follow these steps to configure a project for deployment:

1. To view the SSRS web services URL (uniform resource locator), open the Reporting Services Configuration Manager, one of the utilities installed with SQL Server.

2. Connect to your instance of SSRS as shown in Figure 8-7. If it is not a named instance, then the Report Server Instance will be MSSQLSERVICE.

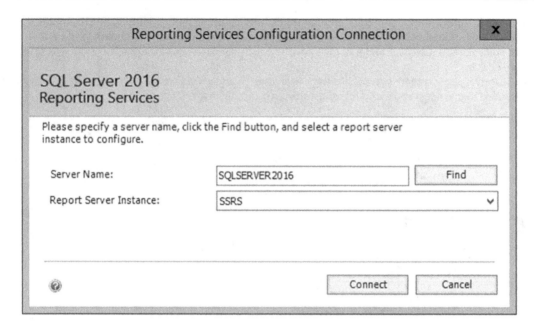

Figure 8-7. The Reporting Services Configuration Connection dialog connecting to a named instance

3. On the Reporting Services Configuration Manager, click Web Service URL on the menu on the left.

4. Record the URL shown in Figure 8-8. This information will be needed in a subsequent step.

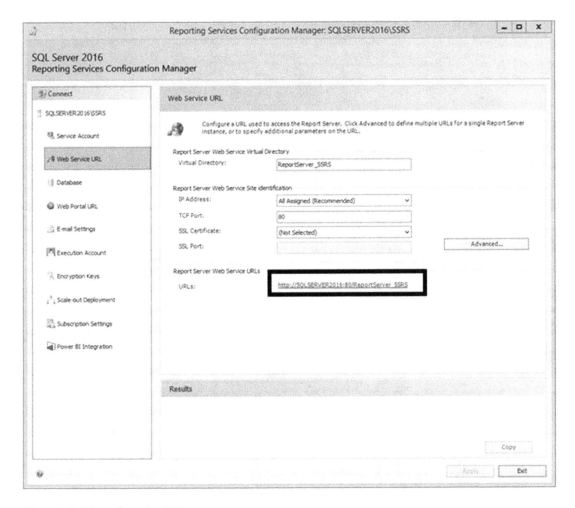

Figure 8-8. *The web service URL*

5. Click Exit to close the Reporting Services Configuration Manager.

6. Launch SSDT.

7. Open the solution created in Chapter 7 by locating it in the File menu under Recent Projects and Solutions. If you did not follow along in Chapter 7, you can find the solution in the Code/Download area of the Apress web site (www.Apress.com) for this book.

8. In the Solution Explorer, right-click the project name, Visual Reports, and select Properties.

9. In the Deployment section, you will see several configuration items with defaults filled in. If it is not already populated, enter the URL that you recorded in step 4 into the TargetServerURL property as shown in Figure 8-9.

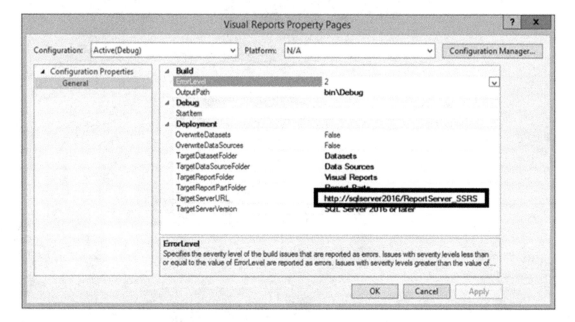

Figure 8-9. *Configure the TargetServerURL*

10. Click OK to save the change. You may have noticed that the port number automatically disappears if it's at the default of 80.

Take a look at Figure 8-10. The OverwriteDatasets and OverwriteDataSources properties are set to False by default. This prevents you from overwriting data sources and datasets that have been deployed. This will come in handy if your data sources are pointing to development or test database servers. When you deploy a project to production, you won't overwrite existing data sources with development connection strings.

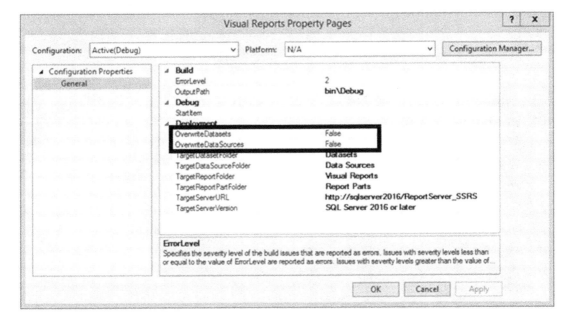

Figure 8-10. *The OverwriteDatasets and OverwriteDataSources properties*

Now that your project is configured, you can deploy the project. Follow these steps:

1. Right-click the project name and select Deploy as shown in Figure 8-11.

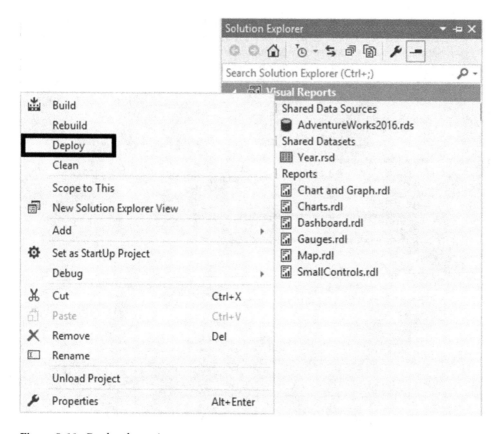

Figure 8-11. *Deploy the project*

■ **Note** If your SSRS installation is local, you may experience permission problems deploying the project and viewing the web portal. See the section "Configuring Local SSRS Settings" in Chapter 1.

---

2. Review the Output window to see if the deployment was successful as shown in Figure 8-12.

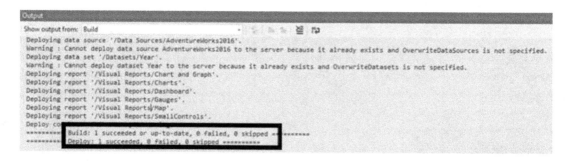

Figure 8-12. *The Output window*

3. Now that the project is deployed, you can view the reports in the web portal. To
    find the URL to use, go back to SSRS Configuration Manager. This time select
    Web Portal URL. To open the web portal, click the URL as shown in Figure 8-13.

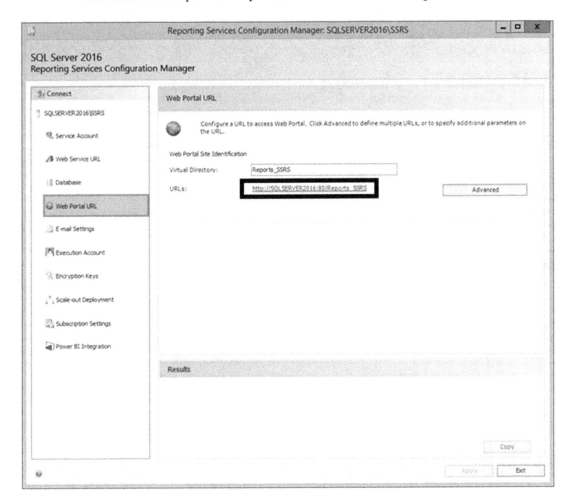

Figure 8-13. *The Report Manager URL*

The Visual Reports folder should now be found in the Home folder of the web portal as shown in Figure 8-14.

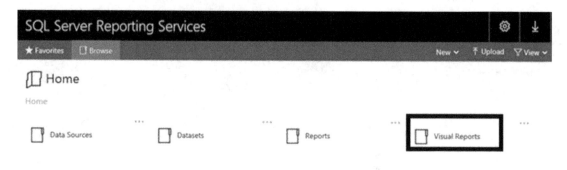

Figure 8-14. *The new folder*

■ **Note**   At the time of the SQL Server 2016 release, the web portal would sometimes fail to run after a restart of the server. If you see a Service Unavailable error, you should stop and start the service by using the Reporting Services Configuration Manager.

If you click the Visual Reports folder, you will see all the reports you created in Chapter 7. You can click a report to view it. Any parameters will be shown across the top. Figure 8-15 shows the Charts report.

Figure 8-15. *The Charts report*

If there is a reason to deploy an individual report, dataset, or data source, you can just right-click the object from the Solution Browser in SSDT and select Deploy.

Just as from SSDT, you can print or export the report. Older versions of SSRS used an ActiveX control for printing which caused many compatibility problems with modern web browsers. Currently the printer icon allows you to save the report as a PDF which can then be printed. The save icon allows you to export the report in several formats including PowerPoint.

Uploading Reports

Instead of deploying from SSDT, you can also publish a report by uploading to the web portal. Follow these steps to upload a report:

1. Launch the web portal.

2. While in the Home folder, create a new folder by clicking New ➤ Folder as shown in Figure 8-16.

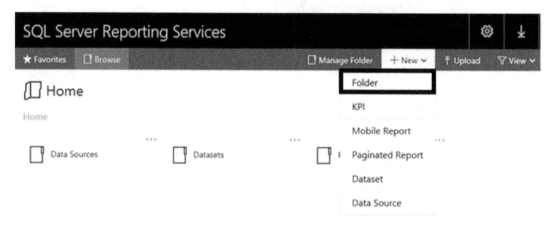

Figure 8-16. *Create a new folder.*

3. Name the folder Upload Example and click Create.

4. Once the folder is created, click it to open it.

5. Click Upload and navigate to a report. The report files have the extension rdl and are found in the project folders as shown in Figure 8-17.

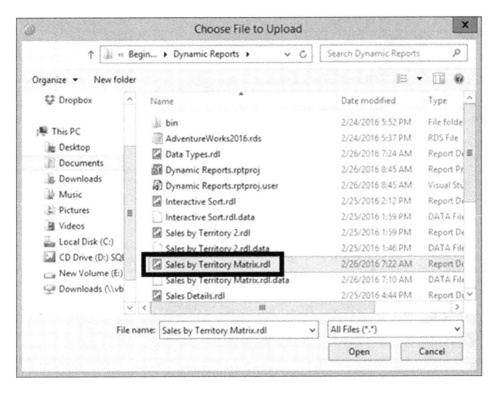

Figure 8-17. *Navigate to a report*

6. Once you select a report, click Open.

7. You should now see the report in the folder. Click the report to view it.

Instead of opening the report, you will see an error message as shown in Figure 8-18.

Home > Upload Example > Sales by Territory Matrix

The report server cannot process the report or shared dataset. The shared data source 'AdventureWorks' for the report server or SharePoint site is not valid. Browse to the server or site and select a shared data source. (rsInvalidDataSourceReference)

Figure 8-18. *The error message*

When you upload a report, it is not automatically linked to the data source. You can easily correct this by following these steps:

1. Click Upload Example in the path to navigate away from the error message.

2. Click the ellipsis next to the report name and select Manage.

3. This opens a page with many options. Select Data Sources. You will see a message about the data source error.

4. Click the ellipsis under Connect to as shown in Figure 8-19.

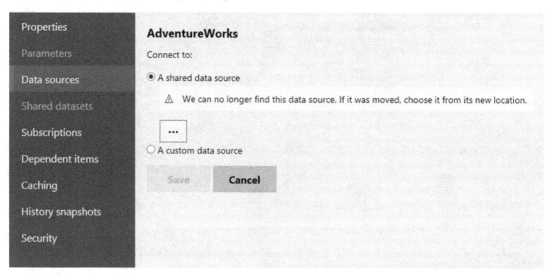

Figure 8-19. The data sources properties

5. This opens up a window at the Home folder. Click Data Sources.

6. Select the AdventureWorks2016 data sources as shown in Figure 8-20.

Data Source

Home > **Data Sources**

Figure 8-20. The Data sources folder

7. You should now see the correctly mapped data source as shown in Figure 8-21.

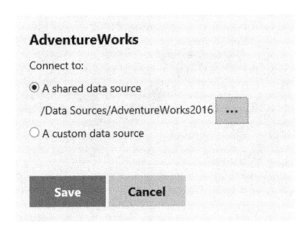

Figure 8-21. *The corrected data source*

8. Click Save.

9. Navigate back to the Upload Example folder and try running the report. It should now run as expected.

Obviously, it is easier to deploy from SSDT, but this information will be helpful if you do ever need to upload an individual report.

Creating Data Sources

The AdventureWorks2016 data source was created automatically when the project was deployed. You can also create data sources manually through the web portal. Follow these steps to create a data source:

1. Launch the web portal.

2. Click the Data Sources folder.

3. Click the ellipsis next to AdventureWorks2016 and then click Manage to open the properties.

4. Select the Connection String property and copy it to the clipboard as shown in Figure 8-22.

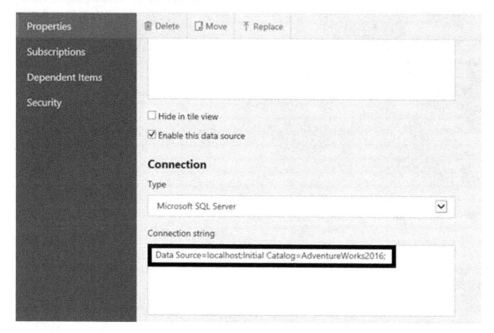

Figure 8-22. *The Connection string property*

5. Scroll down and click Cancel to close the properties.

6. Click New ➤ Data Source to open the New Data Source page.

7. Fill in TestDataSource for the Name.

8. Paste the previously copied connection string into the Connection string property.

9. Scroll down and click Test Connection.

10. Click Create.

11. You should now see both data sources in the folder.

While you may not need to create many data sources through this interface, you will need to use it to manage data sources. Whenever a database is moving to a new server, you will make the change here for the reports. In Chapter 9, you will learn about the Credentials section of the data source properties.

Deploying Report Parts

In addition to data sources, datasets, and reports, you can also deploy report parts. Report parts are the objects that make up reports such as charts, tables, and gauges. The published report parts can then be used with the Report Builder tool that will be covered in Chapter 10.

To publish an object on a report, you must mark individual items for publishing. Follow these steps to deploy the parts of a report:

1. If it's not already open, launch SSDT and the project from Chapter 7.

2. From the Solution Explorer, double-click the Charts report to open it in design view.

3. If the Report menu is not visible, click the design canvas of the report.

4. Select Report ➤ Publish Report Parts which opens a dialog box showing the objects that can be published as shown in Figure 8-23.

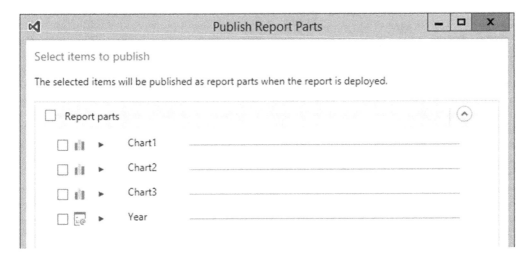

Figure 8-23. *The Publish Report Parts dialog*

5. Check Report Parts to select each item.

6. Click the arrow next to the first chart. This allows you to see a picture of the chart and also fill in a description.

7. Each item has the default name which will not be helpful in the web portal. Change the name of each to match Figure 8-24.

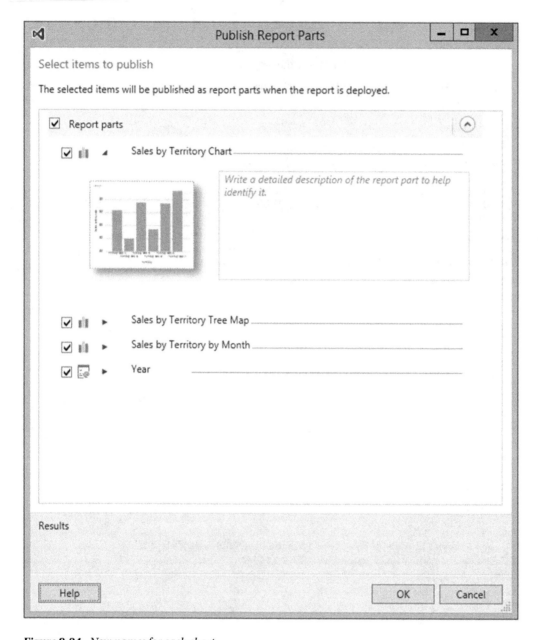

Figure 8-24. *New names for each chart*

8. Click OK.

9. In Solution Explorer, right-click the Charts report and deploy it.

10. Open the web portal.

11. Navigate to the Home folder.

12. You should see a new Report Parts folder.

13. Click the folder to see the published report parts as shown in Figure 8-25. You will use them in Chapter 10.

Figure 8-25. *The Report Parts folder*

Deploying Reports to SharePoint

Installing and configuring SSRS in SharePoint mode is beyond the scope of this book. I do think, however, that if you ever need to deploy reports to an existing SharePoint farm, you can come back to this chapter to learn how to do it. Just like native mode SSRS, you can publish reports either by uploading or from SSDT. This section assumes you are using SharePoint 2013 or later and that the SSRS document library has been configured properly.

Follow these steps to learn how to upload SSRS reports to a SharePoint report library:

1. Launch SharePoint and navigate to the report library. Figure 8-26 shows a typical report folder.

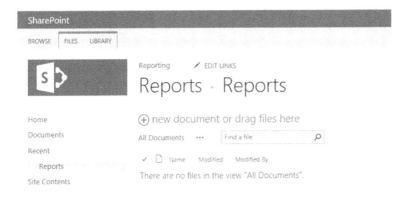

Figure 8-26. *The SharePoint report folder*

2. Click Files.

3. Click Upload Document as shown in Figure 8-27.

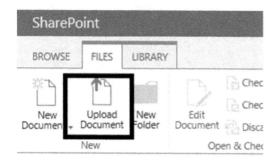

Figure 8-27. *The Upload Document icon*

4. From the Add a Document dialog, navigate to an SSRS report rdl file by clicking Browse.

5. Click OK.

6. On the Reports dialog, select Report Builder Report as the Content Type as shown in Figure 8-28.

Reports - SQL Account Product List.rdl   ✕

EDIT

Save   Cancel   Paste   ✂ Cut   ⬚ Copy   Delete
                                                  Item

Commit   Clipboard   Actions

ⓘ The document was uploaded successfully. Use this form to update the properties of the document.

Content Type	Report Builder Report ▾
	Create a new Report Builder report.
Name *	Product List ✕	.rdl
Title	

Created at 3/16/2016 8:40 PM by ☐ MYDOMAIN\domainadmin
Last modified at 3/16/2016 8:40 PM by ☐ MYDOMAIN\domainadmin   Save   Cancel

Figure 8-28. *The Reports dialog*

7. Click Save. The report should now show up in the folder.

8. Because you uploaded the file, you will need to configure the data source. Click the ellipsis next to the report name.

9. This opens up a menu dialog; click the ellipsis on this dialog.

10. Select Manage Data Sources as shown in Figure 8-29.

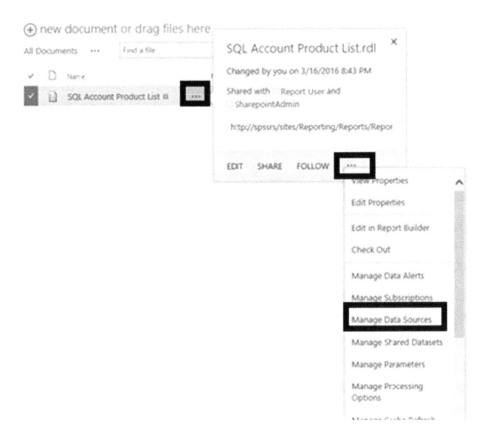

Figure 8-29. *The report properties menu*

11. On the Manage Data Sources page, click the data source name.

12. On the data source's property page, click the ellipsis as shown in Figure 8-30.

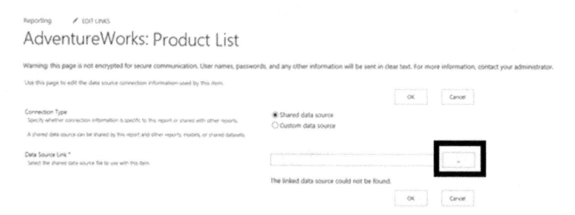

Figure 8-30. The data source properties

13. On the Select an Item dialog, navigate to the correct data source as shown in Figure 8-31. Click the Up button to navigate.

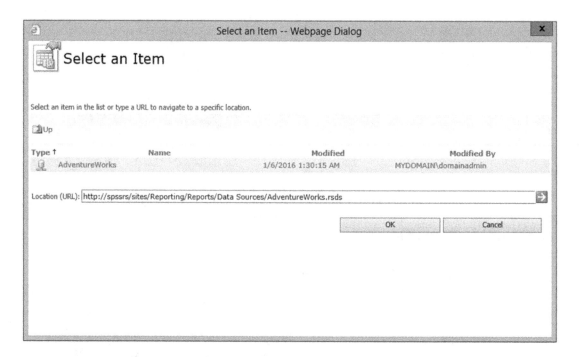

Figure 8-31. The Select an Item dialog

14. Click OK to save the data source name and dismiss the dialog box.

15. Click OK on the data source properties for the report to save the change.

16. Navigate to the report and test it.

You can also upload data source files (rds) or create them. To create a new data source, follow these steps:

1. Launch SharePoint and navigate to the Data Sources folder.

2. Click Files ➤ New Document as shown Figure 8-32.

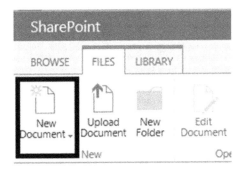

Figure 8-32. *The New Document icon*

3. Select Report Data Source as shown in Figure 8-33.

Figure 8-33. *The Report Data Source menu item*

4. Fill out the properties on the Data Source Properties page as shown in Figure 8-34.

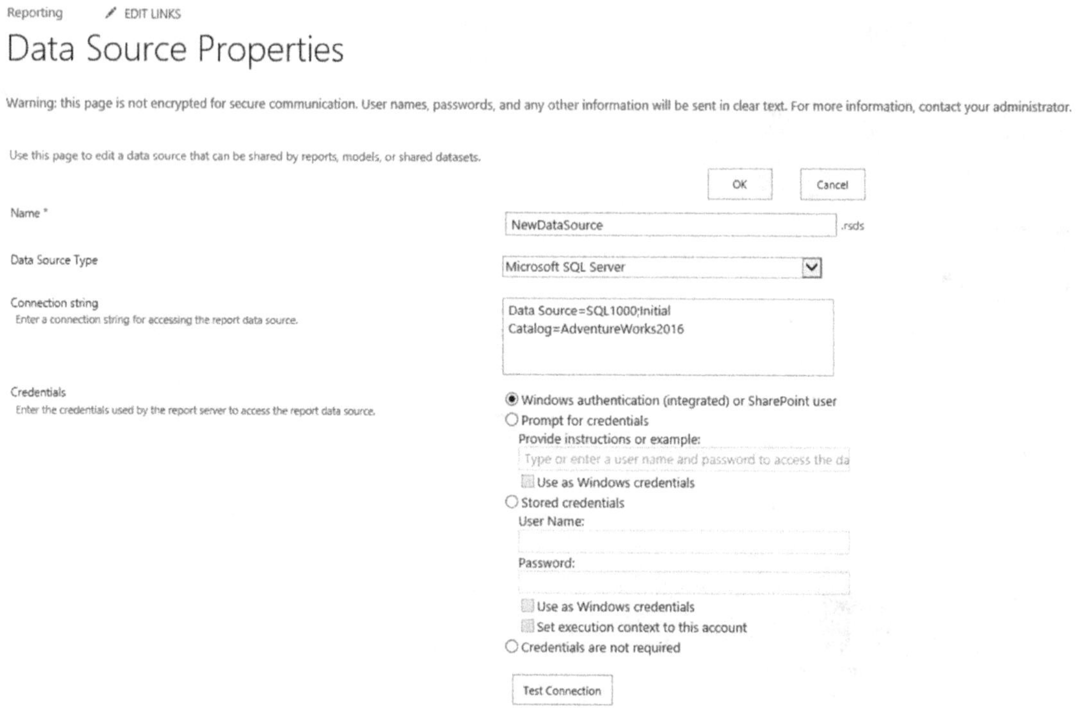

Reporting   ✎ EDIT LINKS

Data Source Properties

Warning: this page is not encrypted for secure communication. User names, passwords, and any other information will be sent in clear text. For more information, contact your administrator.

Use this page to edit a data source that can be shared by reports, models, or shared datasets.

OK   Cancel

Name *   NewDataSource   .rsds

Data Source Type   Microsoft SQL Server ▾

Connection string
Enter a connection string for accessing the report data source.

Data Source=SQL1000;Initial
Catalog=AdventureWorks2016

Credentials
Enter the credentials used by the report server to access the report data source.

◉ Windows authentication (integrated) or SharePoint user
○ Prompt for credentials
  Provide instructions or example:
    Type or enter a user name and password to access the da
    ☐ Use as Windows credentials
○ Stored credentials
  User Name:

  Password:

    ☐ Use as Windows credentials
    ☐ Set execution context to this account
○ Credentials are not required

Test Connection

Figure 8-34. *The Data Source Properties*

5. Click OK to save the new data source.

To deploy reports and other objects from SSDT, you will need to configure the exact paths to the SharePoint library folders for each type of item in the project properties. Otherwise, the process is identical to deploying to a native mode instance. Figure 8-35 shows how to configure for my SharePoint site.

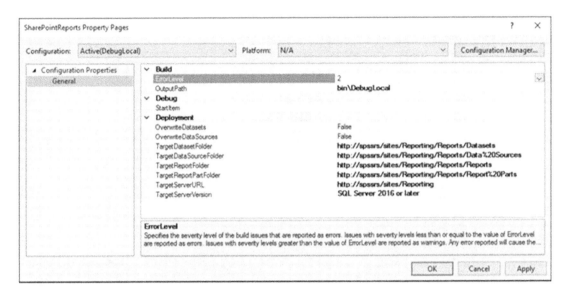

Figure 8-35. *The properties when deploying to SharePoint*

Summary

SSRS reports are not much value unless they are published where the end users can get to them. In this chapter you were introduced to the new web portal. You deployed reports from SSDT and by uploading them. If you have an SSRS instance installed in SharePoint mode, you also saw how to deploy reports to SharePoint.

Whenever data is concerned, security is of utmost importance. In Chapter 9, you will learn about the security aspects of SSRS.

CHAPTER 9

Securing Your Reports

A major data breach is reported on an almost weekly basis. One of the first that caught my attention was the 2012 theft of South Carolina's state taxpayer database. In 2015, hackers stole the electronic records of millions of US federal government employees. A report developer may not need to configure security, but he or she should be able to understand how SQL Server Reporting Services (SSRS) security works to assist the administrator if requested.

When deploying SSRS reports, there are two layers of security to consider: permissions at the data source and within SSRS. In this chapter, you will learn the security features of SSRS and how SSRS interacts with SQL Server security.

Understanding SQL Server Security

Even if an end user is allowed to run a report, the SQL Server instance my not return the requested data. The settings in the data source determine the credentials sent to SQL Server.

Note A thorough review of SQL Server security is beyond the scope of this book. Data for reports can come from many other sources as well, such as Oracle databases and Analysis Services cubes. Consult vendor documentation to understand the security of any particular product.

In the case of SQL Server, two types of accounts will be used to connect: Windows logins and logins defined within the SQL Server instance. Both types of logins are mapped to database users which have specific permissions within the database. Another feature of SQL Server called Contained Databases allows you to create users directly within the database. Figure 9-1 illustrates these concepts.

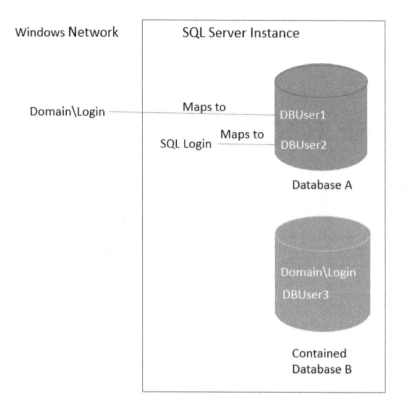

Figure 9-1. *Authentication within SQL Server*

Setting Up an SQL Account

By default, SQL Server will accept only Windows authentication. SQL Server authentication can be enabled during installation, or you can configure it after the fact. To follow along with all of the examples in this chapter, SQL Server authentication must be enabled. If you installed your SQL Server instance locally, you will be able to modify the authentication properties. Otherwise, you will need to work with your database administrator to configure security on a development instance in your network. To change the authentication properties of SQL Server, follow these steps:

1. Launch SQL Server Management Studio (SSMS).

2. Type in the SQL Server name property and Authentication method as shown in Figure 9-2. If you are not sure about these properties, refer to the section "Determining the SQL Server Name" in Chapter 1.

■ **Caution** Modifying the server authentication property requires a restart of SQL Server.

Figure 9-2. *The Connect to Server dialog*

3. Click Connect.

4. In Object Explorer of SSMS, right-click the server name and select Properties.

5. Select the Security page

6. If the Server Authentication is set to Windows Authentication mode, switch it to SQL Server and Windows Authentication mode as shown in Figure 9-3.

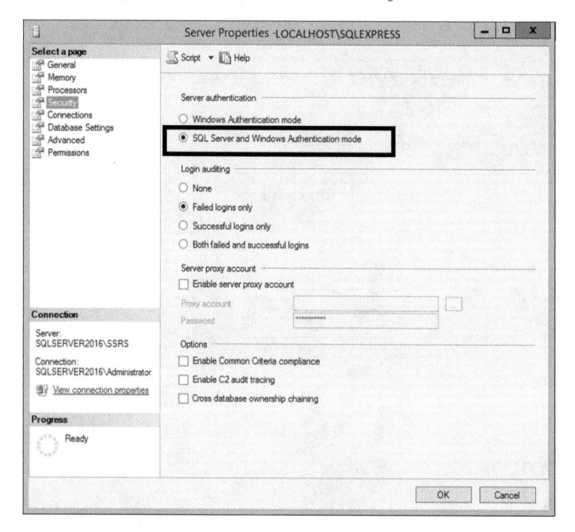

Figure 9-3. SQL Server and Windows Authentication mode

7. Click OK to accept the change.

8. Restart SQL Server by right-clicking the server name and selecting Restart.

Now that SQL Server is configured to accept SQL Server authentication, you will need to create a login and give it permissions. Follow these steps to set up the login:

1. In Object Explorer, expand Security.

2. Right-click Logins and select New Login.

3. This opens the Login – New dialog box.

4. Fill in SQLReportUser for the Login Name.

5. Select SQL Server Authentication.

6. Type in and confirm a password that you will remember.

7. Since this is just for an example, uncheck Enforce password policy as shown in Figure 9-4.

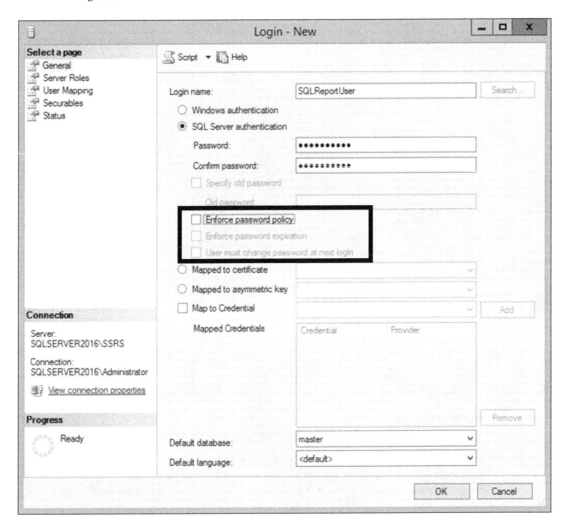

Figure 9-4. *The SQLReportUser account properties*

8. Click OK to create the account.

9. Right-click the account and select Properties.

10. Switch to the User Mapping page.

11. Check next to AdventureWorks2016 as shown in Figure 9-5.

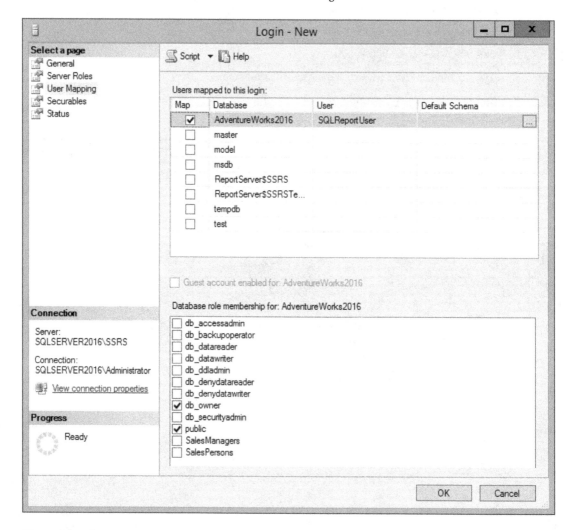

Figure 9-5. *The User Mapping page*

12. Check the db_owner role. In a production database, give only the rights needed. I've noticed that sometimes db_owner is automatically selected for new users in AdventureWorks, but it doesn't stick. Go back and view the properties and reset it if needed.

13. Click OK to save the permissions.

Connecting to SQL Server

In Chapter 3, you learned how to create a data source and most likely used your Windows credentials to connect to SQL Server. In this section, you will create a new data source that uses a SQL Server account. Follow these steps to create a new data source in the project to accept SQL Server credentials:

1. Launch SQL Server Data Tools (SSDT) and open the solution from Chapter 6.

2. Right-click Shared Data Sources and select Add New Data Source.

3. On the General page, set the Name property to AdventureWorks2016SQL as shown in Figure 9-6.

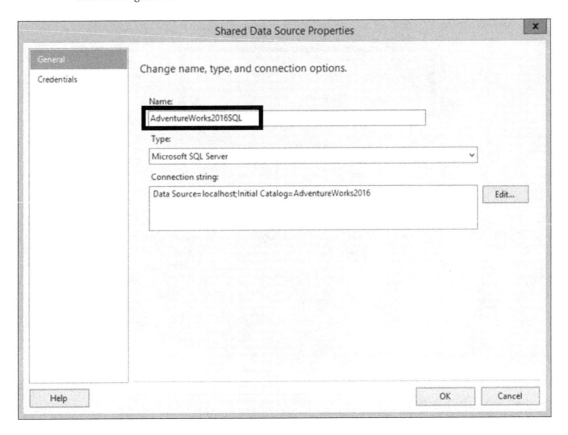

Figure 9-6. *The Name property*

4. As you have done throughout the book, click the Edit button to set the Connection string property.

5. On the Credentials page, select Use this user name and password.

6. Type in the User name and Password from the account you previously created. The Credentials page should look like Figure 9-7.

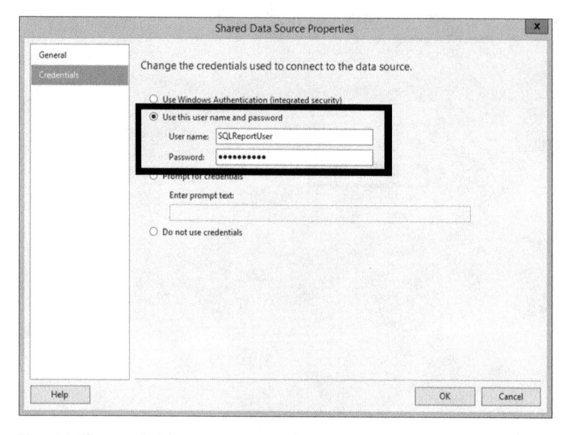

Figure 9-7. *The new credentials*

7. Click OK to save the changes.

8. Double-click the Sales by Territory Matrix report to open it in design view.

9. In the Report Data window, open the properties of the AdventureWorks data source.

10. On the General page, select AdventureWorks2016SQL for the Use shared data source reference property as shown in Figure 9-8.

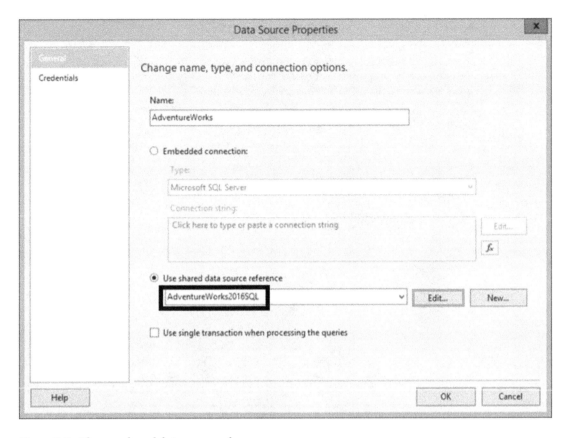

Figure 9-8. *The new shared data source reference*

11. Click OK to save the changes.

12. Preview the report to make sure it works.

Now the report is configured to use a SQL Server account to retrieve the data. When the project is deployed, the end users will connect to the web portal with their Windows accounts. SSRS will then connect to SQL Server with the account credentials that have been stored in the data source properties. Follow these steps to deploy the project and view the new data source in the web portal:

1. In the Solution Explorer, right-click the project name and select Properties.

2. Set the TargetServerURL to your web service URL (uniform resource locator) if it is not already set. See the section "Deploying Reports from SSDT" in Chapter 8 if you are not sure how to do this.

3. Click OK to save the changes.

4. Right-click the project name and select Deploy.

5. If the deployment was successful, launch the web portal. See the section "Deploying Reports from SSDT" in Chapter 8 if you need help with this.

6. Click the Data Sources folder.

7. Click the ellipsis next to the AdventureWorks2016SQL data source and select Manage.

8. This opens the properties of the data source. Scroll down to the Credentials section and view the properties. Notice that you can also specify a hard-coded Windows account. Figure 9-9 shows the Credentials section.

Credentials

Log into the data source

○ As the user viewing the report

◉ Using the following credentials

Type of credentials

Database user name and password ⌄

User name

SQLReportUser

Password

••••••••••••

Figure 9-9. *The data source credentials*

You may be wondering why you do not specify As the user viewing the report all the time. In some cases, it is not possible. Following is a list of possible reasons to save hard-coded credentials in data sources:

- End users or devices may not be part of the Windows domain.
- End users may not be allowed to have direct rights in the database.
- Subscriptions require hard-coded credentials.
- The database server may not reside on the same server as SSRS.

■ **Note** Kerberos delegation must be configured to forward credentials between servers. This is an advanced security topic beyond the scope of this book. To learn more about Kerberos delegation, see the Pluralsight course "Configuring Kerberos for SSRS."

Configuring Site Security

SSRS security is role based. The roles have certain predefined permissions. Adding an account or group to a role allows the account or group to have the permissions defined by the role. Some aspects of security can be configured at the site, while others are based on location or object. In this section, you will learn about site security.

■ **Note** For these examples, I will be using several Windows accounts created on a stand-alone Windows Server 2012 R2. In your environment, you may be running within a domain or on an isolated laptop running one of many end-user versions of Windows. Because there are so many variations possible, I will not walk you through creating the accounts.

I have created an information technology (IT) group and added the following local Windows accounts to the group on my server:

- CIO
- Director
- Manager
- TeamLeader
- TeamMember

At the Site level of the web portal, there are two possible roles: System Administrator and System User. Any accounts or group members in the System Administrator role have full control of the site, including controlling security. The System User property is important for anyone who will be using Report Builder, which you will learn about in Chapter 10. To add an account to a site-level role, follow these steps:

1. Launch the web portal.

2. Click the gear icon and select Site Settings as shown in Figure 9-10

Figure 9-10. *Select Site settings*

3.   This opens the site properties. Click Security.

4.   Click Add group or user as shown in Figure 9-11.

Figure 9-11. *Click Add group or user*

5.   For Group or user, type in CIO.

6.   Select System Administrator as shown in Figure 9-12.

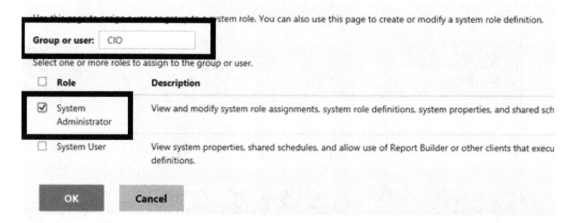

Figure 9-12. *Adding an account*

7.   Click OK to create the role membership.

8.   Repeat the process to add the Manager account to the System User role.

You can also remove accounts from the site-level roles on this page.

Configuring Folder and Report Security

Before you publish the first report in your company, you should plan the folder structure within the web portal to simplify security management. End users can view those folders and reports that they have rights to see. As a best practice, configure security at a folder level only. Security can be configured on individual reports, but that makes long-term management very difficult.

To reach a particular folder, an end user should have rights to the folders above it. By default, folders inherit permissions from the parent folder and reports inherit permissions from the folder in which they are located. Permissions should be set to be more restrictive as you travel down the path. For example, you may want to grant rights to everyone in the domain at the Home folder. Then create folders for each department under Home. Within each department folder, create folders for managers.

There are several roles defined at the folder and object level. Table 9-1 lists the roles and definitions as displayed on the Security page.

Table 9-1. *The Folder- and Object-Level Roles*

Role	Purpose
Browser	May view folders and reports and subscribe to reports.
Content Manager	May manage content in the Report Server. This includes folders, reports, and resources.
My Reports	May publish reports and linked reports; manage folders, reports, and resources in a user's My Reports folder.
Publisher	May publish reports and linked reports to the Report Server.
Report Builder	May view report definitions.

Follow these steps to create and configure the folders for the IT department:

1. In the web portal, navigate to the Home folder. If you are in Site Settings, click SQL Server Reporting Services at the top of the page.

2. Click Manage Folder as shown in Figure 9-13.

Figure 9-13. *The Manage Folder icon*

3. This opens the Security page of the Home folder. Click New Role Assignment.

4. Type in Users. For a domain, this will be the Everyone group.

5. Check Browser as shown in Figure 9-14. This will give all users rights to run any report in the site unless the security in a folder has been overridden.

Figure 9-14. *Add Users to the Browser role*

6. Click OK to save the change.

7. Click the Home link to navigate back to the folders.

8. Click New Folder.

9. Fill in IT for the name and click Create.

10. Click the ellipses next to the new IT folder and select Manage.

11. This opens the folder properties. Click Security.

12. Click Customize security as shown in Figure 9-15.

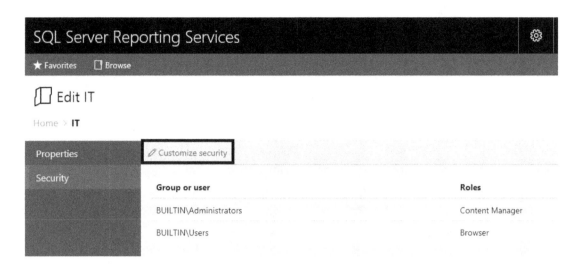

Figure 9-15. *The security page of the folder*

13. This brings up a warning about breaking inheritance from the parent as shown in Figure 9-16. Click OK.

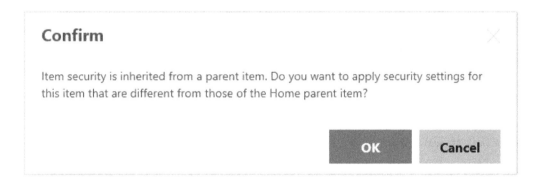

Figure 9-16. *Confirm that security will be different*

14. Click Add group or user.

15. Add the IT group to the folder in the Browser role and click OK.

16. Check the box next to Builtin\Users. If you are working within a domain, remove the Everyone group.

17. Click Delete and OK to confirm.

18. Navigate back to the Home folder.

You now have a folder that can only be viewed by members of the IT group. Continue adding folders and modifying permissions using the skills learned in the previous steps according to Table 9-2 to set up security for the site.

Table 9-2. *The Security for the Site*

Folder	Account or Group	Role Membership
Home/IT	IT	Browser
Home/IT	Manager	Content Manager
Home/IT	Manager	Report Builder
Home/IT/Management	CIO, Director	Browser
Home/IT/Management	Manager	Content Manager
Home/IT/Management	Manager	Report Builder
Home/IT/Management/Confidential	CIO	Content Manager

Obviously, this is a simplified example, but it should illustrate how security should be configured. Figure 9-17 demonstrates how the number of accounts diminishes the further down you go in the path.

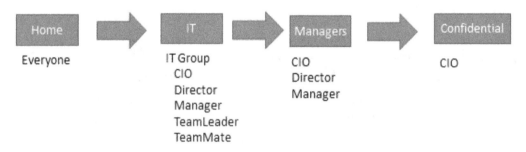

Figure 9-17. *The folder permissions*

To test, run the browser as any of the accounts to make sure that the account can only see the expected folders. Figure 9-18 shows the difference between what the Director and TeamLeader see when each is looking at the IT folder.

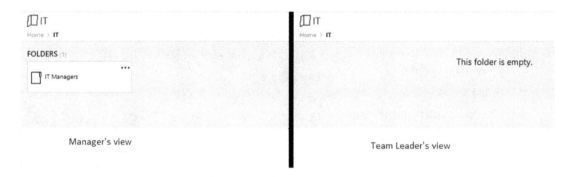

Figure 9-18. *The Manager's and TeamLeaders's views of the folder*

At this point, the Manager account has control of the content of most of the folders and permission to use Report Builder. You will see Report Builder in action in Chapter 10.

Sending Reports Automatically with Subscriptions

One of the nicest features of SSRS is the ability to schedule reports to run and be delivered automatically. The reports may be sent via e-mail or delivered to a network share. When delivering an e-mail report, either the report may be embedded in the e-mail message or a link can be sent.

■ **Note**   Subscriptions are supported with SQL Server Standard Edition and higher.

Because subscription delivery does not involve human intervention, there are two requirements that must be met for a subscription to be created for a particular report. All data sources used by the report must contain hard-coded credentials. Default parameters must be determined and configured in the subscription.

In previous versions of SSRS, an SMTP (Simple Mail Transfer Protocol) server within your domain was required to send reports through e-mail. Starting with SSRS 2016, it is now possible to use an e-mail server outside your domain such as gmail to send reports. To send reports to a network share, the share must be created. A Windows account must have permission to create the report in the share, and the appropriate end users must have permission to open the files. SQL Server Agent must also be running on the server that hosts the ReportServer database for both types of subscriptions.

Before you can send reports through e-mail, the SMTP account settings must be configured. Select the E-mail settings page to fill in the SMTP properties. See the Reporting Services Team blog found at `https://blogs.msdn.microsoft.com/sqlrsteamblog/2016/07/15/deliver-reports-via-emailusing-an-email-server-outside-your-network/` for information about how to configure the settings for several types of services.

It is much simpler to configure the settings for delivery of a report to a network share than via e-mail. The example in this section will cover delivering the report to the share. When you configure a subscription to deliver a report to a share, you will need to specify a File Share Account which has permission to create the file. You can set up one account for the entire SSRS instance, or you can specify an account for individual subscriptions. Figure 9-19 shows the properties for the File Share Account that can be used for subscriptions delivered to network shares.

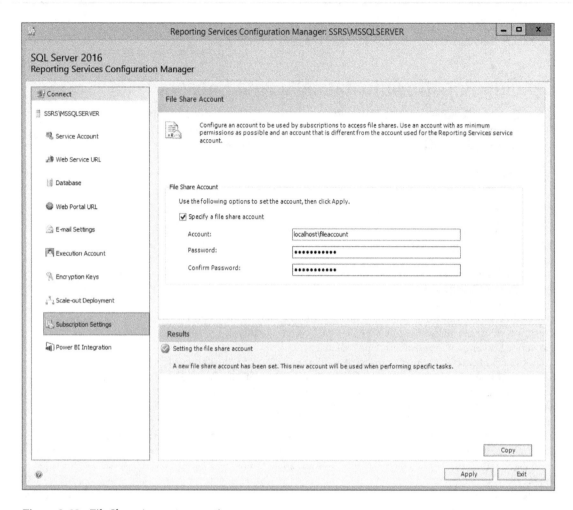

Figure 9-19. *File Share Account properties*

You will also need to create the share where the reports will be delivered. To configure a network share, follow these steps:

1. Navigate to the C:\ drive or another drive on your computer.

2. Create a new folder named Reports.

3. Right-click the folder and select Properties.

4. Select the Sharing tab.

5. Click Share which brings up the File Sharing dialog.

6. Add any accounts and permissions required for the share as shown in Figure 9-20.

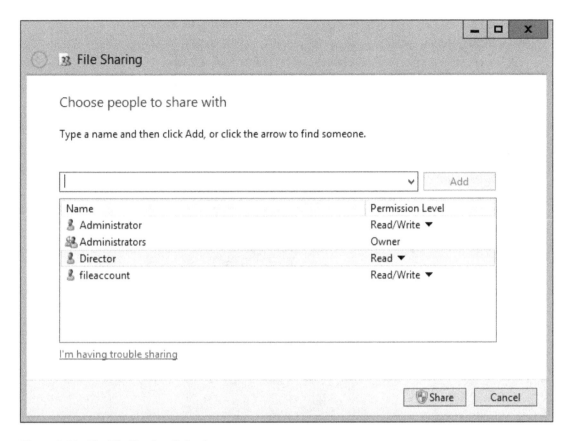

Figure 9-20. *The File Sharing dialog box*

   7.   Click Share, Done, and Close to create the share and close the dialogs.

Once the network share is created, launch SSMS and make sure that SQL Server Agent is running. SQL Server Agent is the service which runs scheduled jobs for SQL Server. SSRS subscriptions are just a type of job as far as SQL Server is concerned.

Now that you have the infrastructure in place, you can create a subscription. Follow these steps to configure the subscription:

   1.   In the web portal, navigate to the Dynamic Reports folder.

   2.   The Sales by Territory Matrix report should be connected to the AdventureWorks2016SQL data source with stored credentials. If not reconfigure the data source.

   3.   Test the Sales by Territory Matrix report.

   4.   Navigate back to the Dynamic Reports folder.

   5.   Click the ellipsis next to Sales by Territory Matrix and select Manage.

   6.   Click Subscriptions.

   7.   Click New Subscription.

8. On the Subscription property page, type Sales by Territory in the Description property.

9. The Owner property should automatically default to your account.

10. For this example, accept the default schedule. For a production subscription, be sure to create a schedule. Figure 9-21 shows the properties so far.

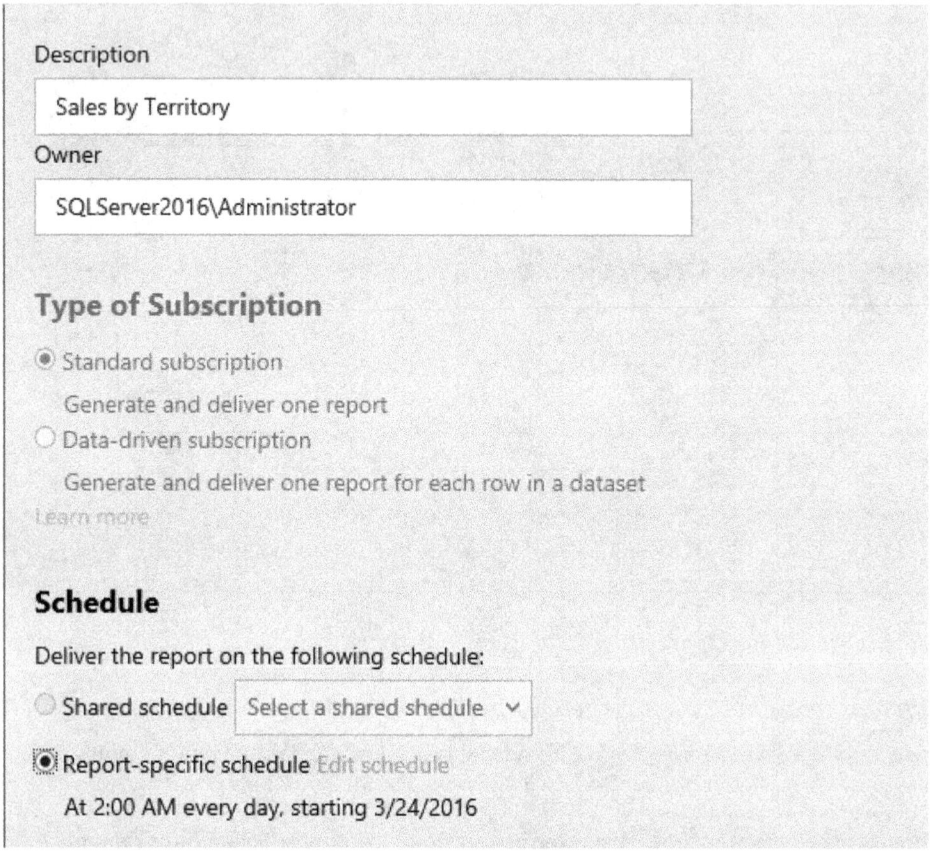

Description

Sales by Territory

Owner

SQLServer2016\Administrator

Type of Subscription

◉ Standard subscription

   Generate and deliver one report

○ Data-driven subscription

   Generate and deliver one report for each row in a dataset

Learn more

Schedule

Deliver the report on the following schedule:

○ Shared schedule   Select a shared shedule ∨

◉ Report-specific schedule Edit schedule

   At 2:00 AM every day, starting 3/24/2016

Figure 9-21. The subscription properties

11. Scroll down to the Destination property and select Windows File Share.

12. Type in the path to the network share.

13. Select PDF for the Render Format.

14. If you wish to override the File Share Account, select Use the following Windows user credentials. Fill in the User Name and Password for the specific account.

15. Modify the Overwrite Options according to your requirements. This section of properties should look similar to Figure 9-22.

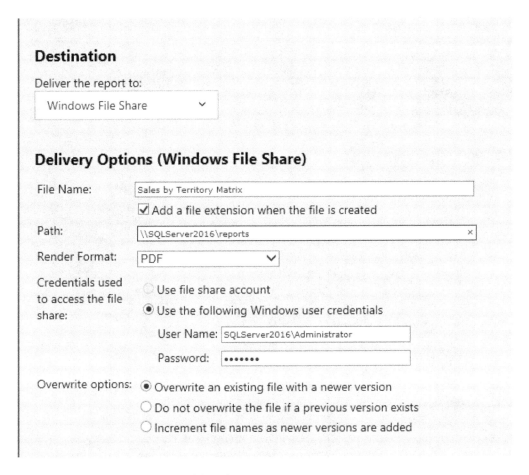

Figure 9-22. *The delivery options of the subscription*

16. Scroll down to the Report Parameters section.

17. Select a value for the Year parameter.

18. Once the Year value is set, the values for Territory will populate. Select a territory. The parameter properties should resemble Figure 9-23.

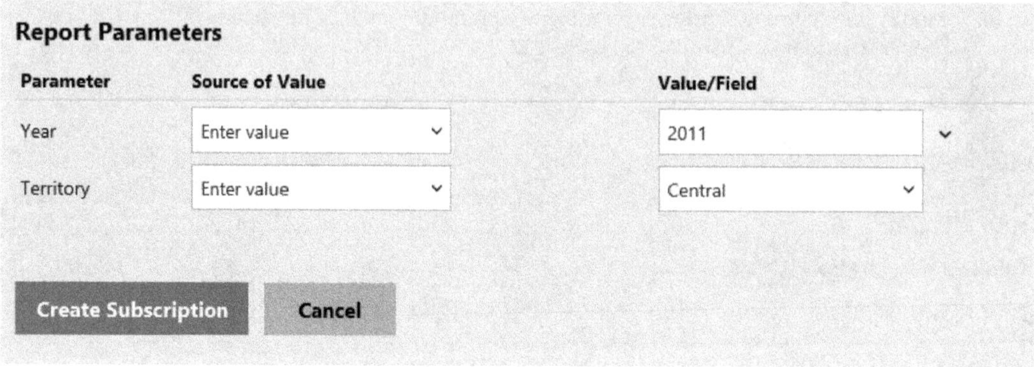

Report Parameters

Parameter	Source of Value	Value/Field
Year	Enter value ⌄	2011 ⌄
Territory	Enter value ⌄	Central ⌄

Create Subscription   **Cancel**

Figure 9-23. *The Report Parameters of the subscription*

19. Click Create Subscription.

20. The subscription should now be visible in the list as shown in Figure 9-24.

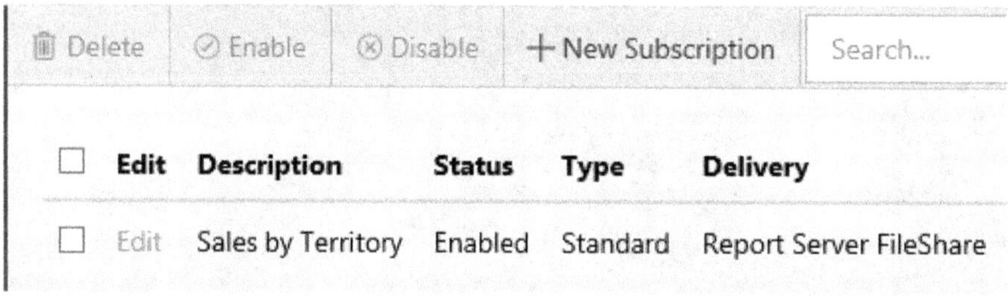

🗑 Delete   ⊘ Enable   ⊗ Disable   ✛ New Subscription   Search...

☐	Edit	Description	Status	Type	Delivery
☐	Edit	Sales by Territory	Enabled	Standard	Report Server FileShare

Figure 9-24. *The new subscription*

A SQL Server Agent job is created for each subscription. Unfortunately, the name of the job is a globally unique identifier (GUID), and you can't tell by looking which job goes with each subscription. To figure out which subscription maps to each job, you will need to run a query like the following in your ReportServer database:

```
SELECT Name AS ReportName, ScheduleID AS JobName, s.[Description]
FROM [Catalog] c
JOIN Subscriptions s ON c.ItemID = s.Report_OID
JOIN ReportSchedule rs ON c.ItemID = rs.ReportID
    AND rs.SubscriptionID = s.SubscriptionID;
```

Figure 9-25 shows the results of the query.

	ReportName	JobName	Description
1	Sales by Territory Matrix	6BC7DDA8-03C5-444C-845E-8312C03C1140	Sales by Territory

Figure 9-25. *The report name mapped to the job name*

To test the subscription, run the job. The job will report success once it kicks off the subscription. It will report success even if the report is not delivered. You can see the results by checking the share for the report or refreshing the Subscriptions page to see the status.

In previous versions of SSRS, you had to navigate to every folder and report to find and manage subscriptions. The web portal has a new feature called My Subscriptions. By clicking the gear in the menu, you will see a link to My Subscriptions. Here you can see the list of subscriptions owned by the account you are using and their locations. You can also enable, disable, or delete your subscriptions. Figure 9-26 shows the link and the list.

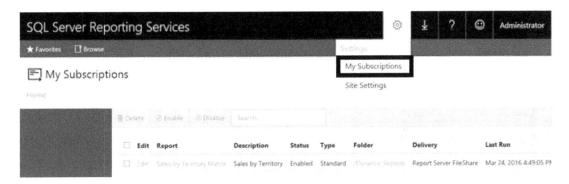

Figure 9-26. *The new My Subscriptions feature*

Securing Delivery

By default, report data travels through the network unencrypted. This may be a security risk in your environment. You can configure SSRS to use the secure socket layer (SSL) protocol. This is the network protocol used to connect securely that you see when buying something or visiting your banking web site. To properly configure SSL, you need to purchase a certificate from a trusted certificate authority.

Once you have obtained a certificate and installed it on the server, it can be configured by launching the Reporting Service Configuration Manager. On the Web Portal URL page, click Advanced. The SSL certificate can be bound to SSRS by clicking Add at the bottom of the Advanced Multiple Web Site Configuration dialog box. Figure 9-27 shows the dialogs you will use to configure SSL.

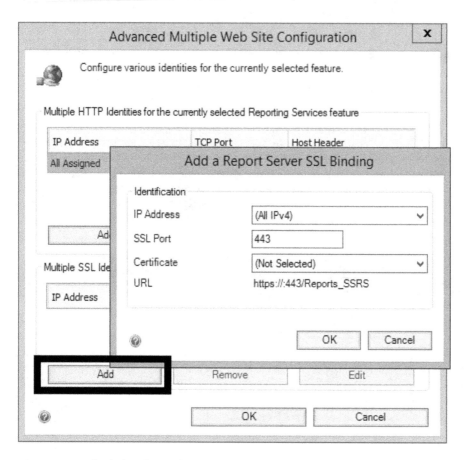

Figure 9-27. *The dialogs for configuring SSL*

Summary

Securing your reports is very important, but it can become very complex and difficult to manage if you don't have a strategy before you start publishing reports. Make sure you figure out the layout of the folders and never configure security on a report, only on a folder.

Security is configured in two layers: at the data source and in SSRS. Predefined server and folder roles make security simpler to manage. You can also configure reports to be delivered automatically through subscriptions.

Chapter 10 covers creating reports directly from the web portal with Report Builder. Two types of reports new with 2016, KPIs and Mobile Reports, will be covered as well.

CHAPTER 10

Creating Self-Service and Mobile Reports

Not long ago, I was playing games at a table with my four-year-old granddaughter and my grandson who was approaching his second birthday. My smartphone was out of my reach when it audibly announced a couple of text messages. My grandson grabbed the device and handed it across to me while saying, "Your phone! Your phone!" When I let him look at photos on the phone, he immediately started swiping. I know he was imitating the behavior of the grownups around him, but today's children are exposed to technology from birth.

The world has changed so much over the past few years. We demand access to our data immediately and from wherever we happen to be. Our work and recreation often blend together as we stay connected constantly to social media and our jobs. Whether good or bad, that is the new reality.

Throughout this book, you have created reports using SQL Server Data Tools (SSDT) within Visual Studio. Visual Studio is a developer tool, but end users often want to create their own reports and dashboards. They are also demanding to keep up with data when outside the company intranet on their tablets and phones. In this chapter, you will learn how to create reports using the end-user tool Report Builder. You will also create mobile reports and key performance indicators (KPIs), two new features of SQL Server 2016 available with Enterprise and Developer Editions.

Using Report Builder

Report Builder has undergone several iterations since it was first released with SQL Server 2005. The current version takes advantage of published datasets and report parts to make building a report much easier than within SSDT. It is, however, possible to start from scratch, building a report in a way that is very similar to using SSDT. Report Builder also contains additional wizards not seen in SSDT. There are still the features you learned to use throughout the book such as the property pages and expressions. Table 10-1 lists the differences between SSDT and Report Builder.

© Kathi Kellenberger 2016

K. Kellenberger, *Beginning SQL Server Reporting Services*, DOI 10.1007/978-1-4842-1990-4_10

Table 10-1. *Differences Between SSDT and Report Builder*

Feature	SSDT	Report Builder
Target audience	Report developers	Power users
Look and feel	Visual Studio	Microsoft Office
Organization	Project based	Individual reports
Deployment	Deploy or upload	Save
Report Parts	Deploy report parts	Consume and deploy report parts

Power users and others in the organization who wish to create their own reports will benefit from having prebuilt report parts, parameter lists, and datasets available to them. That will still create work for the report developers, but it will help ensure that the power users are able to use Report Builder efficiently without needing to be expert T-SQL programmers or to completely understand the source databases.

Chapter 8 described how to deploy reports, including report parts. If you did not follow along during Chapter 8, go back and complete the steps in the sections "Deploying Reports from SSDT" and "Deploying Report Parts." To get started with Report Builder, follow these steps:

1. Launch the web portal.

2. Create a new folder by selecting New ➤ Folder from the menu.

3. Type in Ad Hoc Reports as the name and click Create.

4. Once you have created the folder, click it to navigate inside.

5. Click New ➤ Paginated Report as shown in Figure 10-1.

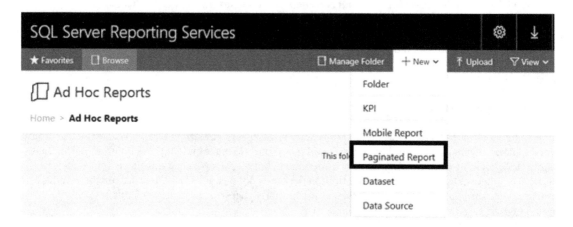

Figure 10-1. *Create a new paginated report*

6. You may see a message asking if you wish to let the browser allow the application to launch. Click Allow.

7. If this is the first time you have launched Report Builder, you will see a message telling you to download and install it as shown in Figure 10-2.

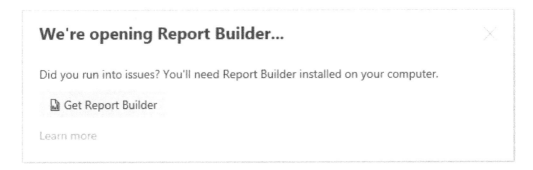

Figure 10-2. *Link to install Report Builder*

8. The Get Report Builder link will take you to a page on Microsoft.com where you can download the app. Follow the instructions on the page. I continued to see the message even after I installed Report Builder, but I just dismissed it.

9. When running the installation wizard, you will be prompted to enter a target server URL (uniform resource locator). Enter the web service URL that you determined in Chapter 8 in the section "Deploying Reports from SSDT." Figure 10-3 shows an example URL.

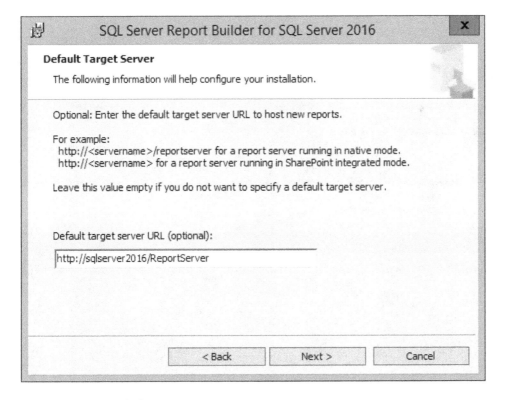

Figure 10-3. *The Default Target Server property*

10. After the installation, you may need to select New ➤ Paginated Report from the menu once again to actually launch the Report Builder tool.

11. Agree to any warnings that may appear.

12. Once Report Builder launches, you will see the New Report or Dataset dialog as shown in Figure 10-4.

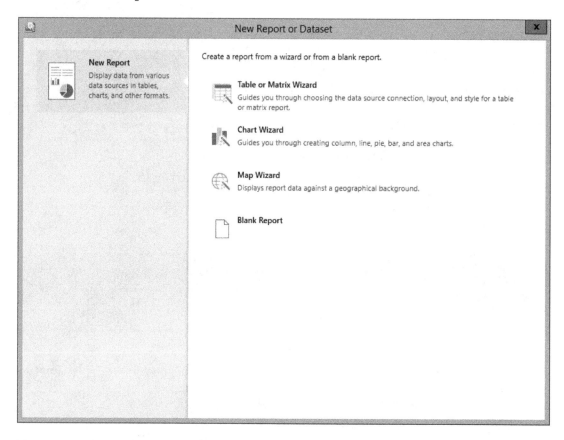

Figure 10-4. The New Report or Dataset dialog

13. Select Blank Report.

14. You will see the report design area as shown in Figure 10-5.

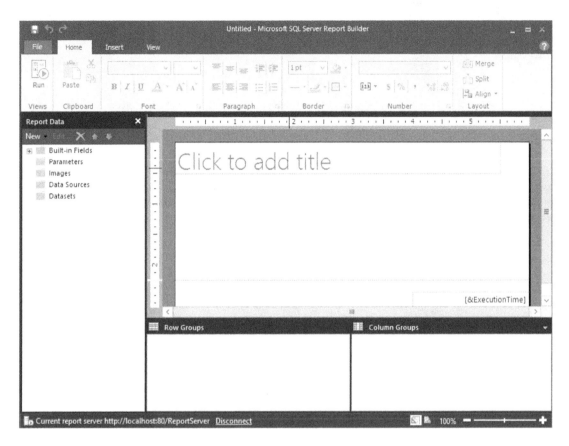

Figure 10-5. *The report design area*

You'll notice that the design area looks very similar to the design area you have seen while developing reports in SSDT. The menu style, however, is more like the ribbon menus found in Microsoft Office. There is no toolbox; you will add controls from the Insert ribbon shown in Figure 10-6.

Figure 10-6. *The Insert ribbon*

The View ribbon shown in Figure 10-7 toggles the visibility of windows in the application.

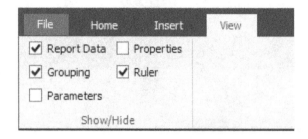

Figure 10-7. *The View ribbon*

You can use the skills you have learned throughout the book to build reports. You can also use the various wizards that come with Report Builder. In this section you will learn to do something that can't be done in SSDT, incorporate existing Report Parts into a report. To use Report Parts, you must be connected to a running SQL Server Reporting Services (SSRS) instance. Follow these steps:

1. From the Insert ribbon, click Report Parts.

2. This opens the Report Part Gallery window on the right side as shown in Figure 10-8.

Figure 10-8. *The Report Part Gallery*

3. By default, you search for report parts based on the name. You can search additional criteria by selecting Add Criteria and selecting an item as shown in Figure 10-9. Select Type.

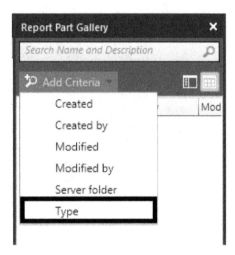

Figure 10-9. *Add Criteria menu*

4. This causes the Type menu to display as shown in Figure 10-10. Select Chart.

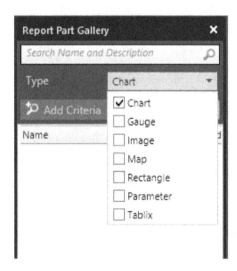

Figure 10-10. *The Type menu*

5. You can type a name in the Search Name and Description box, but if you just click the magnifying glass icon, all the Chart items will display as shown in Figure 10-11. As long as there are not many published parts, this is fine.

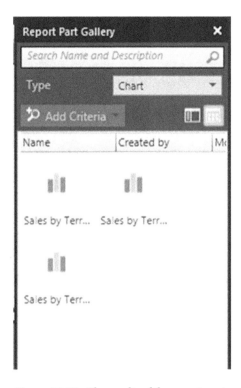

Figure 10-11. *The results of the report part search*

6. To remove the Type, click the Type label and select Remove as shown in Figure 10-12.

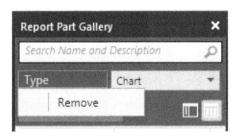

Figure 10-12. *Remove a criteria*

7. Drag one or more of the charts to the report. You will see that the Report Data objects are created automatically as shown in Figure 10-13.

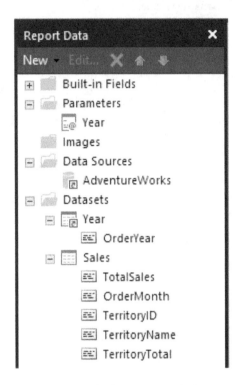

Figure 10-13. *The Report Data window*

8. To view the report inside Report Builder, click the Run icon found on the Home ribbon as shown in Figure 10-14.

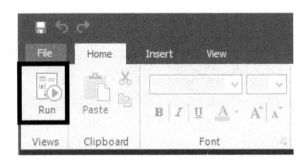

Figure 10-14. *The Run icon*

9. To go back to Design view, click the Design icon in the Run ribbon as shown in Figure 10-15.

Figure 10-15. *The Design icon*

Once the report is complete, you can either save it to your file system or publish it. Follow these steps to publish the report:

1. From the File menu, click Save. This brings up a dialog that looks like a file save dialog as shown in Figure 10-16.

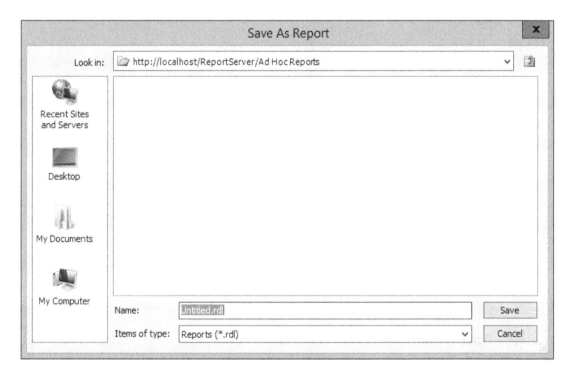

Figure 10-16. *The Save As Report dialog*

2. Type in a name for your report and click Save. By default, it is saved in the folder from where you launched Report Builder. You can also navigate to other folders as needed.

From the dialog shown in Figure 10-16, you can also save the report definition locally. The File menu shown in Figure 10-17 will let you save and open reports from various locations as you might expect. You can also publish report parts that you have created. The Check For Updates item updates report parts that your report uses in case the originally published report part has been changed.

Figure 10-17. *The File menu*

From within the web portal, from the Manage menu on a report, you can launch Report Builder to perform edits as shown in Figure 10-18. You can also create a shortcut to Report Builder or pin it to the taskbar.

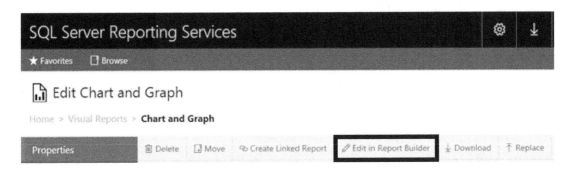

Figure 10-18. *The Edit in Report Builder icon*

Report Builder can be used to create datasets as well as reports. You will create several datasets using Report Builder in subsequent sections of this chapter.

Report Builder is a fantastic tool. In fact, some shops use it exclusively for report development. The only downside that I see is that it does not integrate with source control software. You can use source control, but it will be a manual process.

To give a power user the ability to use Report Builder and to publish reports, make sure they are in the site-wide System Users group and are in the Browser, Publisher, and Report Builder roles in any folders where they should be allowed to create reports.

Creating KPIs

In previous versions of SSRS, you have been able to add KPIs, or key performance indicators, to reports. Starting with 2008 R2, you could add the Indicator control to a Tablix cell. Before that, you could use expressions to dynamically control the color of a cell or to display an image to accomplish the same thing with a bit more work. SSRS 2016 gives you the ability to create and display independent KPIs within the web portal. These KPIs can give immediate information about goals and important metrics at a glance without running a report. They depend on having shared datasets in place.

KPIs can be very simple or quite complex depending on the properties you define. You can display just a number, compare a value to a goal, display a status, display a trend, or display some combination of these.

Follow these steps to create a dataset that will be used in a KPI:

1. In the web portal, navigate to the Datasets folder.

2. Click New ➤ Dataset which will launch Report Builder. If you see a warning about running the program, click Allow. Report Builder will open ready to create the dataset.

3. On the Choose a data source connection or model to create a shared dataset page, select the AdventureWorks2016 data source if it is visible in the dialog.

4. If you do not see the data source you are looking for, click Browse other data sources and navigate to the Data Sources folder to find it as shown in Figure 10-19.

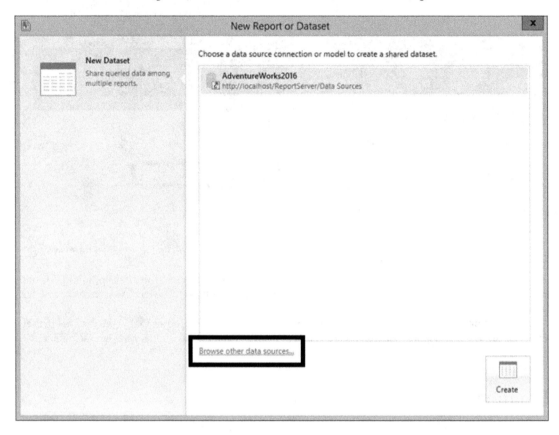

Figure 10-19. *Choose or locate a data source*

5.   Click Create.

6.   This opens the Query Designer where you can build a query by selecting tables and columns as shown in Figure 10-20.

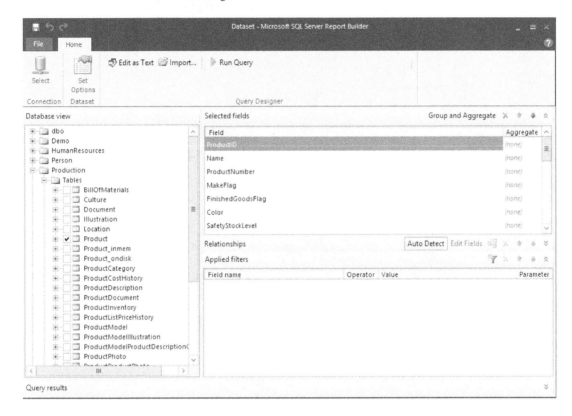

Figure 10-20. *The Query Designer*

7. In this case, you will paste in a query. Click Edit as Text shown in Figure 10-21 to switch from a visual query builder to a text-based query designer window.

Figure 10-21. *Click Edit as Text*

8. Paste in the following query:

```
WITH
Sales AS (
SELECT SUM(TotalDue) AS Sales, MONTH(OrderDate) AS OrderMonth,
    YEAR(OrderDate) AS OrderYear,
    LAG(SUM(TotalDue),12) OVER(ORDER BY YEAR(OrderDate),
        MONTH(OrderDate)) * 1.10 AS Quota
FROM Sales.SalesOrderHeader
GROUP BY MONTH(OrderDate), YEAR(OrderDate)
),
Comparison AS (
SELECT OrderYear, Sales, OrderMonth, Quota,SUM(Sales)
        OVER(PARTITION BY OrderYear)/SUM(Quota)
        OVER(PARTITION BY OrderYear) AS PercentOfGoal
FROM Sales)
SELECT Sales, OrderMonth, Quota,
    CASE WHEN PercentOfGoal >= .98 THEN 1
        WHEN PercentOfGoal > .9 THEN 0
        ELSE -1 END AS Status
FROM Comparison
WHERE OrderYear = 2014;
```

9. Run the query by clicking the exclamation point icon. The results should look like Figure 10-22.

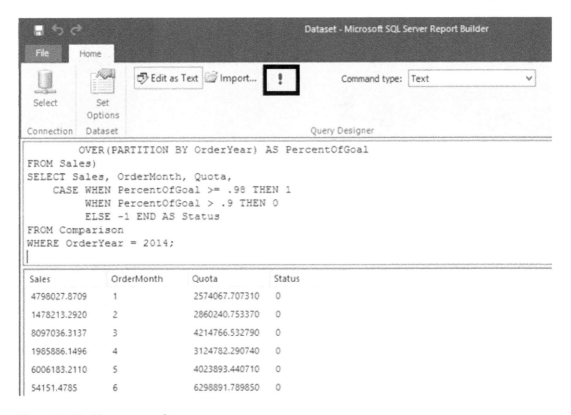

Figure 10-22. *The query results*

10. From the File menu, click Save.

11. Once the Save as Dataset dialog opens, navigate to the Datasets folder.

12. Name the dataset 2014 Sales and click OK.

13. Close Report Builder.

Now that the dataset is published to the web portal, you can create a KPI that uses it. One very interesting aspect of the KPI design page is that you can design how the KPI will look before you connect the data. In fact, when you initially create a KPI, it will be populated with a random value and a visualization. Follow these steps to create the new KPI:

1. Navigate to the Ad Hoc Reports folder or create it if you did not follow along with the section "Using Report Builder."

2. Click New ➤ KPI from the menu.

3. This opens a page shown in Figure 10-23 where you can see all the properties of the KPI including random values already populating it.

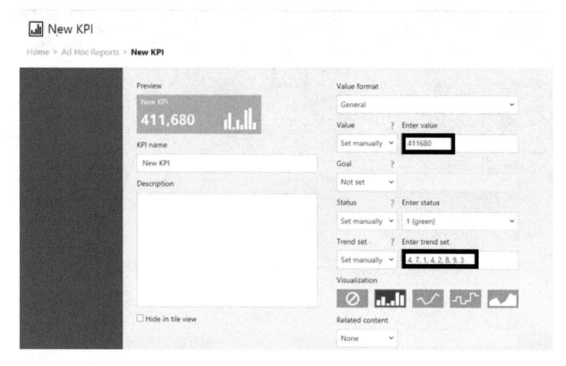

Figure 10-23. *The KPI design page*

4. Change the KPI Name to 2014 Sales.

5. Change the Value format to Abbreviated currency.

6. Change the Value setting to Dataset field.

7. Click the ellipsis under Pick dataset field as shown in Figure 10-24.

Figure 10-24. *Click the ellipsis*

8. This opens the Pick a Dataset dialog. Navigate to the Datasets folder.

9. Select the 2014 Sales dataset as shown in Figure 10-25.

Figure 10-25. *Select the Sales dataset*

10. This opens a dialog that allows you to select the field and aggregation. Select Sum for the Aggregation.

11. Select Sales as the field as shown in Figure 10-26.

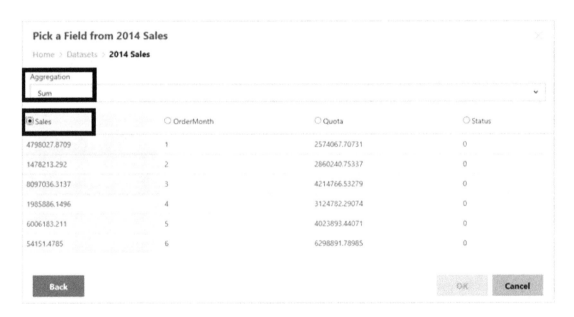

Figure 10-26. *The value properties*

12. Click OK.

13. Follow the same procedure to populate the Goal. Select the Quota field and the Sum aggregation.

14. Use the same dataset for the Status. Select the First aggregation and the Status field.

15. Use the same dataset for the Trend Set. In this case, there is no aggregation. Select the Sales field.

16. Select the Bar Visualization if it is not already selected. The properties should look like Figure 10-27.

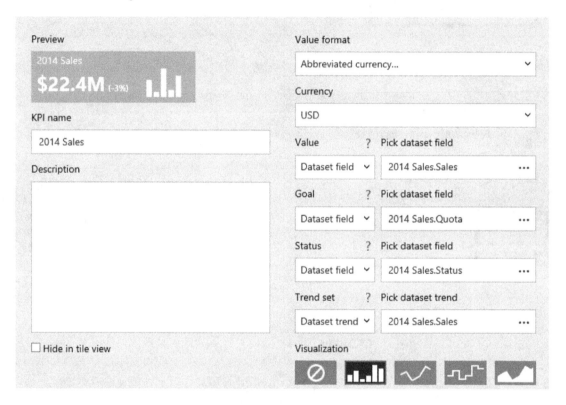

Figure 10-27. *The KPI properties*

17. Click Create at the bottom of the screen.

Now when you navigate to the Ad Hoc Reports folder, you should see the new KPI as shown in Figure 10-28. The sales for 2014 are $22.4 million, and that is 3% under the quota. You can also see the trend over the months in the small chart.

Figure 10-28. *The new KPI*

KPIs can be linked to other reports or web pages. Go back to the editor by clicking the ellipsis inside the rectangle and selecting Manage. To add a link to a paginated report, select Custom URL for the Related content property found at the bottom of the editor. Paste in the URL copied from one of your reports as shown in Figure 10-29. You can also add a link to a mobile report by selecting Mobile Report. In this case, you will browse to the report to fill in the URL.

Figure 10-29. *The Related content properties*

If Related content is populated, the link will be visible when clicking the KPI as shown in Figure 10-30.

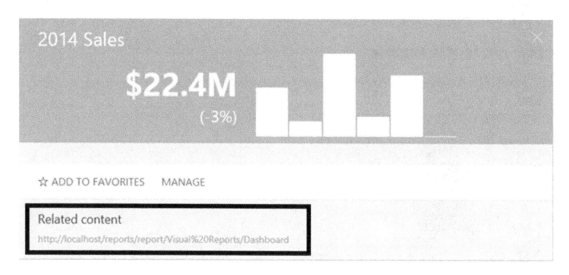

Figure 10-30. *The Related content link*

Creating Mobile Reports

The new mobile reports feature is by far the most anticipated enhancement found with SSRS 2016. Before the feature was announced at the PASS Summit in Seattle in October 2015, I would have guessed this feature would be available only in SharePoint integrated mode. I am thrilled that Microsoft is making this investment in native mode SSRS. In my opinion, native mode is much easier to configure and manage. I'm glad this great feature is available and simple to manage.

When creating a KPI, sample data is created so that you can see how the KPI will look before connecting it to data. The same is true for the mobile reports. To get started you will need to download and install the new Mobile Report Publisher application. You will find the link by clicking the down arrow and selecting Mobile Report Publisher as shown in Figure 10-31.

Figure 10-31. *The Mobile Report Publisher link*

Download and install the application according to the instructions found on the linked page. Once it is installed, you can launch it from the web portal or create a shortcut.

■ **Note**   At the release of SQL Server 2016, the Mobile Report Publisher also required a patch for a Visual C++ component. If that update is not in place, you will be prompted during the installation. Be sure to download the x86 version.

Just like the KPIs, you will need to create shared datasets to build mobile reports. Mobile reports will also work with data imported from Excel. The downside to Excel data is that it must be refreshed manually.

■ **Note**   Starting with 2016, you can store Excel spreadsheets in SSRS. At least with the version available at the time of this writing, the files stored in SSRS could not be used for the mobile reports.

To get started, you will build the mobile report with simulated data by following these steps:

1.   From the web portal, click New ➤ Mobile Report.

2.   OK any warnings about running the application. To avoid this message in the future, create a shortcut to the app instead of launching through the web portal.

3.   You should now see the SQL Server Mobile Report Publisher where you will build the mobile report as shown in Figure 10-32.

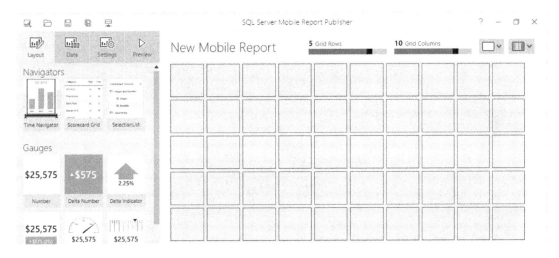

Figure 10-32. The mobile report design canvas

4. From the left side, drag a Selection list to the top and left part of the grid. Drag the right side so that it expands to cover four squares. The Selection list will be used to filter the report.

5. Drag a Half-donut under Selection list. Expand it so that it covers an area of two-by-two squares.

6. Drag a Gradient heat map next to the Half-donut. Expand it so that it also covers a two-by-two square area.

7. Fill the bottom squares with a Category chart. The layout should look like Figure 10-33.

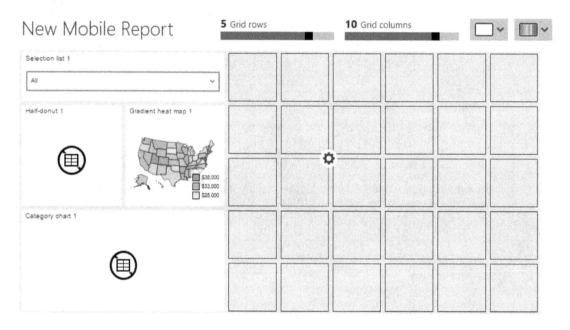

Figure 10-33. *The mobile report layout*

8. Click the Preview button. The report should look like Figure 10-34.

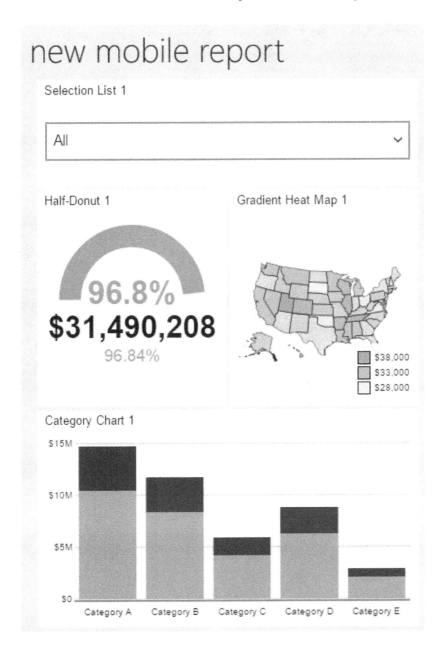

Figure 10-34. *The mobile report preview*

9. From Selection List 1, select one of the items. You will see the data change automatically.

10. Click the title of Category Chart 1. The chart will expand to fill the screen.

11. If you click one of the columns and hold down the cursor, the data for the column will display as shown in Figure 10-35.

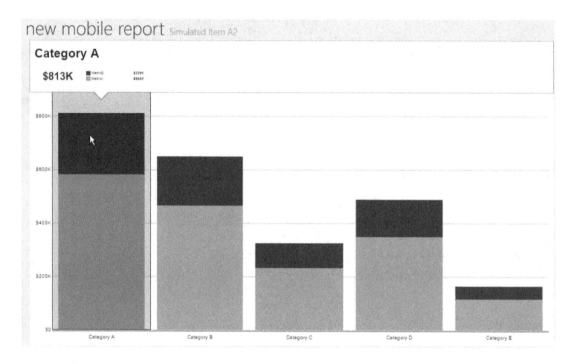

Figure 10-35. *The data for one column*

12. Click the Category Chart title to go back to the report.

13. Click the back arrow to the left of the report title to go to design view.

I hope you are impressed at this point at how you easily you can prototype a report and how interactive it is. The next step is to give each report item a title. Table 10-2 shows the title for each item.

Table 10-2. *The Report Item Titles*

Item	Title
Selection List	Year
Half-Donut	Quota
Gradient Heat Map	US Sales by State
Category Chart	Sales by Month

Follow these steps to set the titles:

1. Inside the Mobile Report Publisher design view, make sure that the Layout view is selected.

2. Select an item from the design grid.

3. Fill in the Title found in the Visual properties section as shown in Figure 10-36.

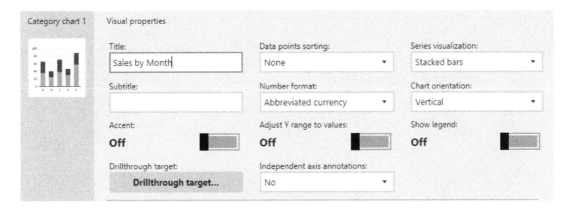

Figure 10-36. *The Title property*

4. Repeat the process for each item.

5. To make sure that you do not lose your work, save the report definition on your local machine as an rsmobile file.

At this point, you have a design, but it is connected to simulated data. The next step is to connect the report to the SSRS server. Follow these steps.

1. Click the Server Connections icon shown in Figure 10-37.

Figure 10-37. *The Server Connections icon*

2. Fill in the Server address as shown in Figure 10-38.

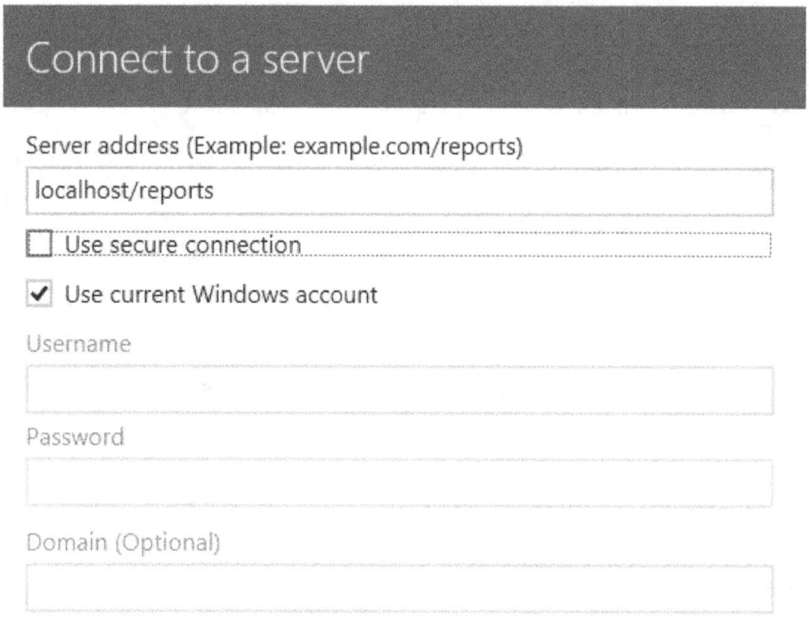

Connect to a server

Server address (Example: example.com/reports)

localhost/reports

☐ Use secure connection

✔ Use current Windows account

Username

Password

Domain (Optional)

Figure 10-38. *The Connect to a server properties*

3. Unless you are using secure socket layer (SSL), uncheck Use secure connection. If SSRS is installed on your local workstation, you are probably not using SSL.

4. Click Connect. The Mobile Report Publisher will remember the connection.

The next step is to add datasets to the web portal that will be used to populate the mobile report instead of using simulated data. Follow these steps to create the datasets:

1. Minimize the Mobile Report Publisher.

2. From the web portal, select New ➤ Dataset to launch Report Builder.

3. Create a new shared dataset named SalesByState using this query and the AdventureWorks2016 data source. The section "Using Report Builder" has instructions on creating a dataset if you need help.

```
SELECT SUM(TotalDue) AS Sales, vS.CountryRegionName, vS.StateProvinceName,
       YEAR(OrderDate) AS OrderYear
FROM Sales.SalesOrderHeader AS SOH
JOIN Person.BusinessEntityAddress AS BEA
    ON BEA.BusinessEntityID = SOH.CustomerID
JOIN Person.Address AS A ON BEA.AddressID = A.AddressID
JOIN Person.vStateProvinceCountryRegion AS vS
    ON A.StateProvinceID = vS.StateProvinceID
WHERE CountryRegionName = 'United States'
GROUP BY A.City, vS.CountryRegionName, vS.StateProvinceName,
       YEAR(OrderDate);
```

4. Save the new dataset in the Datasets folder.

5. Create a dataset named Sales with the following query and save it to the Datasets folder:

```
WITH
Sales AS (
SELECT SUM(TotalDue) AS Sales, MONTH(OrderDate) AS OrderMonth,
    YEAR(OrderDate) AS OrderYear,
    LAG(SUM(TotalDue),12,600000) OVER(ORDER BY YEAR(OrderDate),
        MONTH(OrderDate)) * 1.10 AS Quota
FROM Sales.SalesOrderHeader
GROUP BY MONTH(OrderDate), YEAR(OrderDate)
),
Comparison AS (
SELECT OrderYear, Sales, OrderMonth, Quota,
    Sales/Quota AS PercentOfGoal
FROM Sales)
SELECT OrderYear, Sales, 'M' + RIGHT('0' + CAST(OrderMonth AS VARCHAR(2)),2) AS
OrderMonth, Quota,
    PercentOfGoal, DATEFROMPARTS(OrderYear,OrderMonth, 1) AS OrderDate,
    CASE WHEN PercentOfGoal >= 0.98 THEN 1
        WHEN PercentOfGoal >= 0.9 THEN 0
        ELSE -1 END AS Status
FROM Comparison;
```

6. Close Report Builder.

You should now have the two datasets available for use in the new mobile report. You will connect the datasets to the report objects by following these steps:

1. Maximize the Mobile Report Publisher. Open the rsmobile file you just created if it is not already open.

2. Switch to the Data view. You will see a simulated table connected to each object as shown in Figure 10-39.

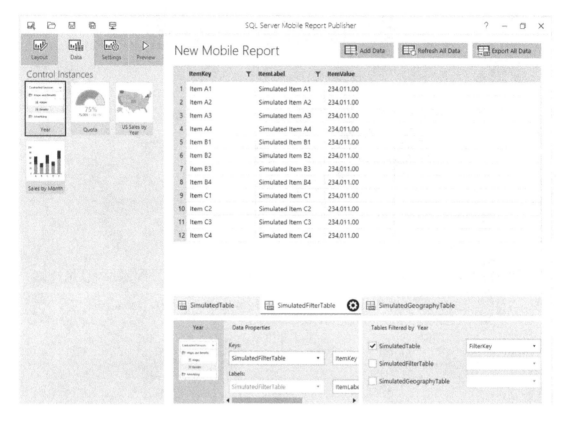

Figure 10-39. The Data view

3. Click Add Data as shown in Figure 10-40.

Figure 10-40. Click Add Data

4. You will have the choice to import from Excel or Report Server as shown in Figure 10-41. Select Report Server.

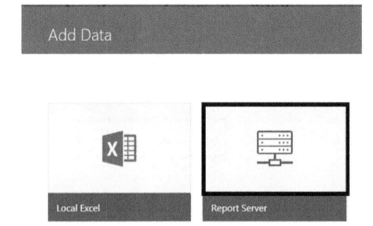

Figure 10-41. *Select Report Server*

5. On the Add Data from Server screen shown in Figure 10-42, select your SSRS instance.

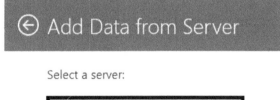

Figure 10-42. *Select the server*

6. You will then see all the folders found directly under the web portal Home folder. Click Datasets.

7. You should now see a list of all the datasets that have been created as shown in Figure 10-43. Click SalesByState.

⊕ Add Data from Server

/Datasets

▦ 2014 Sales	▦ Sales	▦ **SalesByState**
▦ Territory	▦ Year	

Figure 10-43. The datasets

8. Now the SalesByState dataset will be displayed in the Mobile Report Publisher data page instead of the simulated data.

9. Repeat the process to add the Sales dataset.

10. While in Data view, select the Year control.

11. In the Data Properties of the Year control, select Sales for the Keys property and OrderYear for the field next to Keys.

12. Select OrderYear for the field next to Labels. The Data Properties should look like Figure 10-44.

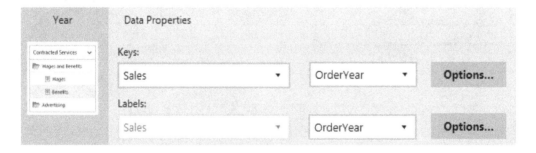

Figure 10-44. The Data Properties of the Year control

13. On the right, you will see the section Filter these datasets when a selection is made. Check both SalesByState and Sales.

14. Select the OrderYear field for each as shown in Figure 10-45.

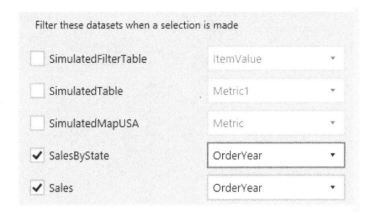

Figure 10-45. *The tables filtered by year*

15. Switch to the Layout view. Make sure that the Year list is still selected and turn off Allow select all as shown in Figure 10-46. This will force the control to allow only one value at a time, not all of them.

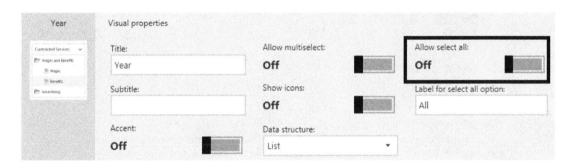

Figure 10-46. *Turn off Allow select all*

16. Switch back to Data view.

17. Select the Quota control. For the Main Value, fill in the Sales dataset and Sales Field.

18. For the Comparison Value, fill in the Sales dataset and Quota field.

19. Click one of the Options buttons, you should see that the data is filtered by the Year control and that the data will be summed as shown in Figure 10-47.

Figure 10-47. *The data options for Quota*

20. Click Cancel.

21. For the US Sales by State control, select the SalesByState dataset and StateProvinceName field for the Keys property. The Keys property should be filtered by Year if you check the Options. If it is not filtered, be sure to select it here and click Done.

22. Select the Sales field for the Values property. The US Sales by State control data properties should look like Figure 10-48.

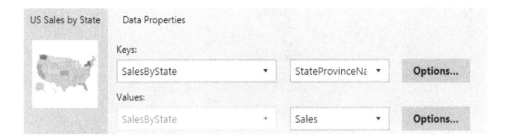

Figure 10-48. *The US Sales by State data properties*

23. For the Sales by Month control, select the Sales dataset and Order Month for the Series name field properties.

24. The Main series field is Sales. This control is also filtered by Year, so be sure to confirm that. The data properties should look like Figure 10-49.

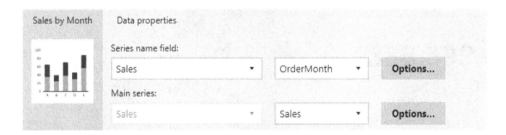

Figure 10-49. *The Sales by Month data properties*

25. Now that you have the controls connected to actual data, you can remove the simulated tables. Click the gear on the simulated table names and click Remove.

The report name is still set as the default; you can click the name in the design view or click the Settings page to change the Report Title property. Set the title to Sales by Month and State. When you preview the report, it should look similar to Figure 10-50.

Figure 10-50. The populated report

26. Switch back to design view by clicking the back arrow.

There are three possible views for a mobile report: Master, Tablet, and Phone. The initial work to select the controls must be done in the Master view. You can then design Tablet and Phone reports based on the Master report. In the upper right-hand corner, switch from Master to Phone as shown in Figure 10-51.

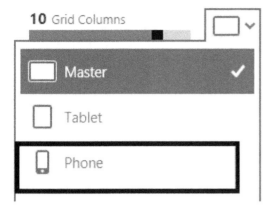

Figure 10-51. Switch to Phone

Changing to Phone view displays the design grid as a smaller size with no controls. On the left, you will see all of the controls that were configured in the Master view. You will drag in and resize the controls you wish to see in the Phone view. Figure 10-52 shows a possible layout for the phone in Preview mode.

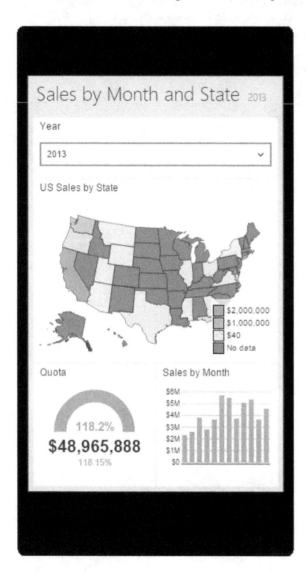

Figure 10-52. Phone preview

While you are viewing the report, try out the dynamic functionality. It still works! To publish the report, you will just save to the SSRS server by following these steps:

1. Switch back to design view.

2. Click the Save button.

3. Select Save to Server.

4. On the Save mobile report as screen, the name will be filled in along with the server name. Click Browse as shown in Figure 10-53.

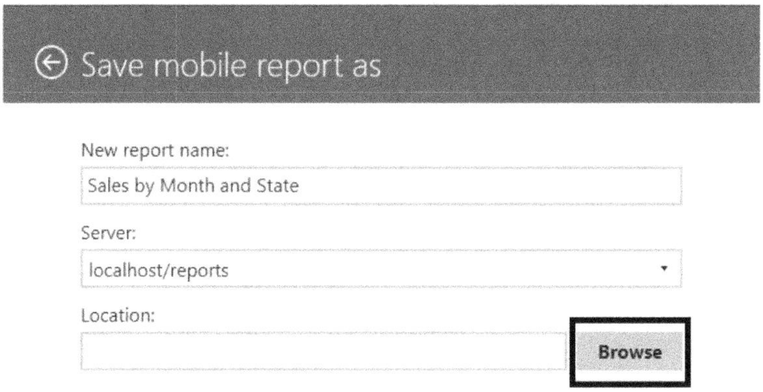

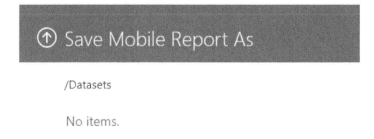

Figure 10-53. *The Save mobile report as screen*

5. Click the up arrow as shown in Figure 10-54 to navigate to the correct folder location if needed.

Figure 10-54. *Click the up arrow*

6. Navigate to the Ad Hoc Reports folder and click Choose Folder.

7. Click Save to publish the report.

Now when you navigate to the Ad Hoc Reports folder in the web portal, you can run the report. If you resize the browser, the report will resize as well as shown in Figure 10-55.

Figure 10-55. *The report automatically adjusts*

Each control also has a Drill-Through Target property that you can set to navigate to another mobile report or to any URL. The new applications for running the reports on mobile devices are Power BI for iOS, Power BI for Android, and Power BI for Windows.

Summary

With the new Mobile Reports feature available with Enterprise Edition, Microsoft has released a product that will deliver data anytime, anywhere, and to many devices. Building the reports is easy to do, and interactivity is built in without special coding. Report Builder has been around since 2005. It has improved considerably since then as a tool to enable power users to build their own reports and dashboards. The new KPIs will also provide information to executives with just a glance and a small amount of development time.

This book is meant for beginners, and there is more to learn. Chapter 11 talks about advanced topics you may wish to explore as you gain experience.

CHAPTER 11

Where to Go from Here

Writing a book meant for beginners in any technology has its own special challenges. What topics are appropriate and which ones are out of scope? How can enough material be included so that the reader gets a good foundation without making the book too large and expensive? Within every single chapter, I thought about the questions someone new to SQL Server Reporting Services (SSRS) might ask from both a developer and an administrator point of view. I tried to explain things in a simple way and cover enough to get you started. I didn't mention some features just to avoid adding unneeded complexity. My hope is that this book inspires you to learn more, and that it just marks the beginning of your journey.

As I wrote Chapter 10, I was already thinking about what this chapter would cover. I realized that I could break the advanced information into five areas: installation, architecture and configuration, development, administration, and integration with other reporting tools. In this chapter, without going too deep into the details, I will tell you about some advanced features in each of these areas.

SSRS Installation

In Chapter 1, you learned how to install SSRS on your own computer in native mode so that you could learn how to deploy and manage reports. The chapter did not cover installing SSRS in SharePoint integrated mode. Over the past few releases of both SQL Server and SharePoint, installing and configuring SSRS in SharePoint mode have become easier, but there are still several steps. You also need a SharePoint farm in place, and many beginners will not have that available to them.

There are a couple of advantages when running SSRS in SharePoint integrated mode. You can add the development of reports to the workflow functionality built into SharePoint. For example, you could require that new reports be approved and that there will be an e-mail notification when changes are made. SharePoint also has document versions, something that you won't find with the web portal.

There are two features that are available with Sharepoint mode but not Native mode. One is Power View reports. These reports are intended for end-user ad hoc reporting using drag and drop. The reports are based on predefined data models, and the person creating the report can explore the data while he or she builds the report. The report is very interactive. By clicking a bar in a chart, for example, the entire report is filtered. Visualizations are quickly changed to different types with just a click. Figure 11-1 is a picture of a Power View report in design view that I created from a predefined data model and about 30 clicks of the mouse. The visualizations reflect actual data.

© Kathi Kellenberger 2016
K. Kellenberger, *Beginning SQL Server Reporting Services*, DOI 10.1007/978-1-4842-1990-4_11

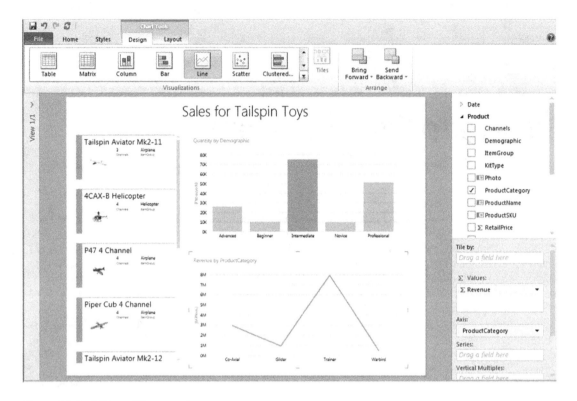

Figure 11-1. *A Power View report*

Another feature only available in reports deployed to SharePoint is data alerts. Based on criteria defined in the data alert, an e-mail will fire when the data of a report changes. This way, the report can be viewed only when there is an important change in the data.

If you like the new web portal but you wish you could customize its look and feel, you are in luck. Starting with SSRS 2016, you can take advantage of the custom branding feature to tailor the look of the web portal. Figure 11-2 is an image of an Xbox-branded web portal from the SSRS Team Blog found at `https://blogs.msdn.microsoft.com/sqlrsteamblog/2016/03/20/how-to-create-a-custom-brand-package-for-reporting-services-with-sql-server-2016/`.

Figure 11-2. *A custom web portal*

Architecture and Configuration

While you have been learning SSRS, you probably have everything you need installed on one computer: a SQL Server instance with a source database and the Reporting Services databases, Visual Studio with SQL Server Data Tools (SSDT), and an SSRS instance. This is the simplest configuration possible. Most of the time, the developer may have only Visual Studio with SSDT installed locally while the SSRS and SQL Server instances are hosted on other servers. Even more complex than this is the scale-out configuration where the SSRS instance is deployed over two or more servers. This configuration allows a larger number of reports to run at any given time. Figure 11-3 shows how the scaled-out architecture looks.

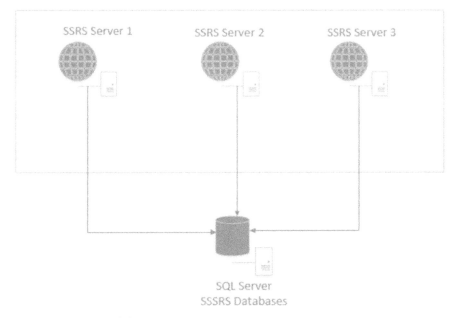

Figure 11-3. *The scaled-out architecture*

Even with the typical SSRS deployment, there are several settings that you may need to change for your particular situation that are not available through the Reporting Services Configuration Manager. Be sure to consult the online documentation about settings you may need to configure in the rsreportserver.config file found in the SQL Server section of Program Files.

Development

Several chapters of this book were devoted to developing reports. You learned about grouping, visualizations, expressions, and more. I once saw a demo where a map visualization was used to grow a virtual flower based on a value from a database. Obviously, the sky is the limit when it comes to designing reports, especially for the more creative among us.

Almost every property of every object can be controlled dynamically with expressions. This book demonstrated how to create a handful of expressions. I encourage you to learn about all the functions that were not used in the book. Some developers add custom functions or even custom assemblies written in C# to their reports.

If you are not that comfortable with writing code, another interesting exercise is creating a template with your company's logo, colors, and footer. You create the report and then save the rdl file in this folder with a descriptive name:

```
C:\Program Files (x86)\Microsoft Visual Studio 4.0\Common7\IDE\PrivateAssemblies\
ProjectItems\ReportProject
```

Then, when you add a new report to a project, the template will be one of the choices and will have all the properties defined in the template by default.

Administration

In Chapters 8 and 9, you learned about deploying reports and controlling the security. You also learned how to set up a subscription. There are several properties of deployed reports that you did not learn about such as caching and snapshots.

Caching is a feature used to save and reuse reports where the data will not frequently change to improve performance. For example, imagine that several people will run a report based on the previous month's sales. The data will not change, so you could take advantage of caching to save the report so that it runs quickly the second and subsequent times it is run.

I once worked on a project that compared data migrated from an old system to data from a new system. A business analyst would select values from a number of parameters on the report and then go back and forth between reports drilling down and viewing the data in different ways. Some of the required queries were painstakingly slow, and there was nothing we could do to improve them. The analysts were spending more time waiting than actually viewing the data. Since the data was not changing that often, I set up caching with a 30-minute expiration. The analysts had to put up with a slow query just the first time it ran with a given set of parameters. This made the solution workable and the analysts were able to do their jobs. Figure 11-4 shows the Caching feature.

Edit Sales by Territory Matrix

Home > Dynamic Reports > **Sales by Territory Matrix**

Properties	○ Always run this report with the most recent data
Parameters	● Cache copies of this report and use them when available
Data Sources	○ Always run this report against pregenerated snapshots
Subscriptions	**Cache Expiration**
Dependent Items	○ Cache expires after `30` Minutes
Caching	● Cache expires on a schedule
History Snapshots	○ Shared schedule `Select a shared shedule ∨`
Security	● Report-specific schedule Edit schedule
	At 2:00 AM every day, starting 3/31/2016
	Cache Refresh Plans
	ⓘ You can manage cache refresh plans after you enable caching by clicking Apply on this page.

Description	Last Run	Status

Apply

Figure 11-4. *The Caching feature*

The other interesting feature is History Snapshots. History Snapshots are similar to subscriptions in that default parameter values must be defined and the data source must store the credentials. You can create a snapshot manually or set up a schedule. Either way, you can go back to view the snapshot later and the report will show the data as it looked at the time the snapshot was taken.

Integration

Since the introduction of SSRS, there has been a Report Manager and now a web portal for hosting the reports. Eventually, integration with SharePoint was added. Developers have embedded reports within applications with a report viewer control. You can also use a special URL (uniform resource locator) to display a report outside the web portal. If you add `rs:Embed=true` to the end of the report's URL from the web portal, it will display in a browser without the web portal heading and menus.

One icon that you may not have noticed, shown in Figure 11-5, from the report menu allows you to export a data feed. The data feed can be used as a data source for Power Pivot, the advanced Excel feature. Power Pivot can be used to combine large amounts of data from one or more sources. The resulting workbook can be used as the basis of Power View reports. Not only is SSRS a way to view data, it can also be the source of data for another reporting tool.

Home > Dynamic Reports > **Sales Summary**

Year   2013

Territory Northwest

|◁ ⊙ 1 of 1 ⊙ ▷| ↻   100%   🖫 ⌄   🖶   [   ] Find | Next 🖫

Figure 11-5. *The Export to Data Feed icon*

SSRS is one of several reporting tools from Microsoft including the time-honored Excel and the new cloud solution, Power BI. SSRS is considered the on-premises reporting solution, especially for paginated reports. With the 2016 release of SSRS, however, SSRS knows no boundaries. SSRS integrates with the cloud solutions Power BI and SharePoint Web, and you have seen the new mobile reports that can be delivered to mobile devices. Over the past few years, some people have predicted the demise of SSRS. With the 2016 revamp, SSRS is alive and well and ready for the future.

Summary

SQL Server Reporting Services is a wonderful topic to learn about. Many people get their start in information technology (IT) as report developers. For many careers, both IT and otherwise, report development is seen as a desired skill. I often talk to people who work in non-IT roles such as human resources who need to learn T-SQL and SSRS for their jobs. In the IT department where I used to work, I was called on to create reports from our help desk ticketing system and our network management tools. Even within IT, we needed to see data from our own systems.

This book is an introduction to the popular reporting tool, SQL Server Reporting Services. This chapter reviewed some of the advanced topics you may want to consider as next steps in your learning. I hope you enjoyed learning about SSRS. I certainly enjoyed writing about it!

Index

Get the eBook for only $5!

Why limit yourself?

Now you can take the weightless companion with you wherever you go and access your content on your PC, phone, tablet, or reader.

Since you've purchased this print book, we're happy to offer you the eBook in all 3 formats for just $5.

Convenient and fully searchable, the PDF version enables you to easily find and copy code—or perform examples by quickly toggling between instructions and applications. The MOBI format is ideal for your Kindle, while the ePUB can be utilized on a variety of mobile devices.

To learn more, go to www.apress.com/companion or contact support@apress.com.

All Apress eBooks are subject to copyright. All rights are reserved by the Publisher, whether the whole or part of the material is concerned, specifically the rights of translation, reprinting, reuse of illustrations, recitation, broadcasting, reproduction on microfilms or in any other physical way, and transmission or information storage and retrieval, electronic adaptation, computer software, or by similar or dissimilar methodology now known or hereafter developed. Exempted from this legal reservation are brief excerpts in connection with reviews or scholarly analysis or material supplied specifically for the purpose of being entered and executed on a computer system, for exclusive use by the purchaser of the work. Duplication of this publication or parts thereof is permitted only under the provisions of the Copyright Law of the Publisher's location, in its current version, and permission for use must always be obtained from Springer. Permissions for use may be obtained through RightsLink at the Copyright Clearance Center. Violations are liable to prosecution under the respective Copyright Law.

CPSIA information can be obtained
at www.ICGtesting.com
Printed in the USA
LVHW05s1753050918
589236LV00008B/182/P